EXPLORING
BRITAIN'S
HIDDEN WORLD

A Natural History of Seabed Habitats

Keith Hiscock

WILD
NATURE
PRESS

To my children, Sally and Peter,
who had to tolerate my many absences on fieldwork
and occasional abandonment while Dad went diving.

To my grandchildren, Maisie and Archie,
who I hope will grow up to love the sea and want to protect it.

Published in 2018 by
Wild Nature Press
Winson House, Church Road
Plympton St Maurice, Plymouth PL7 1NH

A catalogue record for this book is available from the British Library.

ISBN 978-0-9955673-4-4

Printed and bound in Slovenia on behalf of Latitude Press

10 9 8 7 6 5 4 3 2 1

www.wildnaturepress.com

Contents

Foreword

You can wait a lifetime for a book like this. Actually, I've had to wait much of Keith Hiscock's lifetime – long may he prosper! I well remember poring over the pages of seashore guides in my teens as a budding marine scientist. With their cartoon renderings of seaweed, anemone and snail, these books were constant companions as I trudged slippery shores and peered into rock pools. They hinted at deeper mysteries further offshore, beyond reach. There was no guide available for this other world; certainly none that went beyond the mere naming of species. I wish I had possessed a copy of *Exploring Britain's Hidden World* then.

This book is a unique distillation of knowledge and insight built up over a long and storied career. Superlatives are easy to use but rarely fully justified. In this case, however, I feel complete confidence in saying that nobody knows Britain's underwater world as well as Keith Hiscock. His remarkable understanding of this country's marine life transcends the encyclopaedic knowledge of a taxonomist or a museum curator, being based on thousands of hours spent underwater observing, at first hand, the creatures that live there.

Keith's years of research laid the foundations for the habitat scheme used to classify the rich variety of conditions found in European seas. He was an early adopter of internet technology, publishing detailed profiles of thousands of marine species for anyone with a computer connection – a facility that I have used many times. The detail and accuracy of these accounts was peerless, but I longed for a book too. One that I could browse from the comfort of an armchair, just for the pleasure of happening upon something unusual and unsought. This, at last, is that book.

The beautifully illustrated pages invite the reader to dip in again and again, and at each return, to come away with something wonderful, new and unexpected. Britain's seas are as sublime and beautiful as any far-flung regions of the planet featured in television documentaries, but by comparison, they are neglected and little-known. This book is a key to the hidden world that surrounds these islands. So let it peel back the veiled surface and reveal to you the joys that lie beneath.

Callum Roberts is Professor of Marine Conservation at the University of York,
and author of *Ocean of Life: How our Seas are Changing*.

Place names mentioned in the text.

Preface

This book provides a foundation of knowledge for those interested in the natural history of the shallow seabed and demonstrates the marvellous variety of marine life out of sight around Great Britain. The undersea world is fascinating, beautiful and often fragile. The seabed is home to species that are essential to maintaining the richness, health and productivity of our oceans and whose very presence enriches our lives. If seabed habitats and species are to be protected and enjoyed, we need to know what they are and where they are and we need to identify our top conservation priorities.

The area covered is the seabed beyond low water and shallower than 100m off England, Scotland and Wales (Great Britain), not including the Isle of Man, the Channel Islands or Ireland.

The book is the culmination of 50 years of research by the author to better understand where different subtidal seabed habitats occur and how their associated marine life has come to exist. That quest draws on a rich vein of knowledge obtained by many naturalists, scientists and divers who, for almost 200 years, have described seabed communities and sought to understand how they are structured and how they function (how they *work*). Building on that firm foundation, it brings the reader up to date with the latest information about where seabed habitats and their associated communities can be found today, and looks forward to the ways in which we will be exploring our seabed heritage in the future. It is intended to be enjoyed, to inspire and, hopefully, to surprise.

USING THIS BOOK

One of the great differences in publishing a book in 2018 compared with just a couple of years before, is that lists of source references are becoming redundant. With a very few clues in the text, the reader can search the Internet to find the relevant reference and often the actual scientific paper. So, there are no references.

Technical terms are kept to a minimum and are explained wherever possible in the text. There is a glossary of frequently used terms at the end of the book.

The names of habitats and their associated communities of species (biotopes) are given at the end of relevant captions. For the non-specialist reader, ignore them – even experienced scientists struggle with the alphabetic codes. Enter a code into a web search to obtain a full description and distribution map.

For place names mentioned in the text, see the map on page 6.

The scientific names of species change occasionally. The names used in this book have been checked against those listed in the World Register of Marine Species and were correct in December 2017.

Readers can dip into the parts of the book they find interesting and useful. Although the images should leave a lasting memory of what the seabed habitats and species around Britain look like, the sections on how seabed habitats are *shaped* by the environment and how species and habitats are distributed according to environmental conditions help to explain why things are where they are. The historical accounts are important: without understanding the history of study, we will not see current knowledge in context and may even repeat what has already been done. The chapter entitled 'Protecting what we have' will, I hope, make readers think about what really matters – what we should cherish and protect for the future. Good work has already been done to look after seabed habitats and species, but there is much more to do!

Although this book can extend your knowledge of seabed biology, get out there and study it for yourself if you can: that is the only way to fully appreciate and understand ecology. This book is meant to inspire, and to encourage you to put that inspiration into practice. Contribute your own observations of species, of their behaviour and of any changes, to enhance our understanding of the hidden world of the seabed.

For the past 50 years or so, scuba diving has enabled a greater appreciation and understanding of life in Britain's shallow seas. Shallow kelp forest with Dead Men's Fingers and Jewel Anemones at the Eddystone reefs south of Plymouth.

Exploration and discovery: the first 150 years

EARLY DAYS

The full title that I have given this volume includes the words, 'a natural history of seabed habitats'. Our observations provide the foundation of what we know about seabed habitats and their associated communities of species. It was in the 19th century that 'naturalists' or 'natural historians' were beginning not only to document the species found in the seas around Britain, but to make sense of the patterns they were seeing.

From the early part of the 19th century, naturalists were cataloguing the marine life present around the coasts of Britain. Sampling was most often by the use of a dredge. The drawing is from William H. Harvey's *The Sea-Side Book* (1857).

NATURALISTS USING THE DREDGE.

Sampling and description of the biology of subtidal areas has been mainly through the use of a dredge, first mentioned by George Montagu in a paper delivered to the Linnaean Society in 1804. The Irish naturalist Robert Ball was credited as the 'inventor and improver of the naturalists' dredge' in his election as Fellow of the Royal Society in 1857. Although not a great user of the dredge, it was the foremost Victorian marine naturalist Philip Henry Gosse who gave us the most eloquent description of its use.

In September 1865, Philip Henry Gosse wrote (in *A Year at the Shore*):

> Yes; I'm glad I have got a Ball's dredge; and this fair autumnal morning I mean to use it; to go out with honest Harvey, and scrape the stony sea-bottom in the offing yonder. It is a nice portable affair, that one hand can manage; eighteen inches by four and a half are the dimensions of the frame ; the scraping lips are double, one on each edge, so that, however the dredge falls on the bottom, it is sure to scrape. …… Up comes the wet line under Tom's strong muscular pulling, and as it leaves his hands, we coil it snug in the bows of the boat. Dimly appears the dredge some yards below the surface, and now it comes to light, and is fairly lifted aboard. "Tis mortal heavy!" Well it may be, for here is a pretty cargo of huge, rough stones, great oyster-shells, and I know not what. Bright scarlets and crimsons and yellows I discern, and many a twinkling movement among the chaos raises our expectations of something good.

Philip Henry Gosse (1810–1888).

The dredge used to repeat Gosse's 1865 sampling in Torbay. Marine Biological Association Laboratory Steward Peter Rendle is pictured preparing the dredge.

Left: unusually dense clusters of serpulid worms from Torbay in 1865. From *A Year at the Shore* by Philip Henry Gosse.

Right: serpulid worms from Torbay in 2011.

One hundred and forty-six years later, I returned to the area that Gosse had dredged in Torbay with a nearly identical dredge from the ship's store at the Marine Biological Association in Plymouth. Using the dredge together with diving and with the help of the Torbay branch of the British Sub-Aqua Club, we accounted for most of the species that Gosse recorded and identified many more. Unusually dense colonies, rarely seen in British waters, of the worm *Serpula vermicularis* were present both in 1865 and in 2011.

 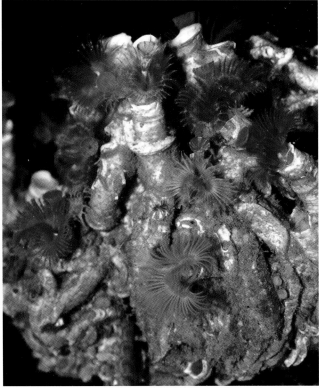

Edward Forbes (1815–1854).

Edward Forbes, in his short career, did much to describe communities sampled by the dredge and to develop and encourage this means of sampling the seabed. It was as a schoolboy of 15 that Forbes started dredging off the Isle of Man, and later, in 1835, he published his findings in the *Magazine of Natural History*. In 1840 the British Association formed a Committee for 'the investigation of the marine zoology of Great Britain by means of the dredge'. In that year a grant of £50 was awarded for that purpose (of which only £15 was spent, in part because of the 'state of the weather, which prevented dredging in the open sea during a great part of the summer'). Further grants followed, and in 1851 Forbes reported on the records of over 140 dredging excursions. Perhaps the most successful programme of dredging was off Shetland, reported by J. G. Jeffries in 1869, which produced both species lists and a comparison of the fauna of the Shetland Islands with that of other parts of the British Isles. However, only a few of the collecting expeditions resulted in the description of assemblages of species from particular locations, and might therefore have contributed to an understanding of what we now describe as communities or biotopes. What those studies did do was to lay the foundation for what is now known as 'biogeography': the distribution of species and habitats according to their geographical location. The map showing the distributional limits of various species and published in 1858 by Forbes (posthumously) is remarkable, and those boundaries are the same today.

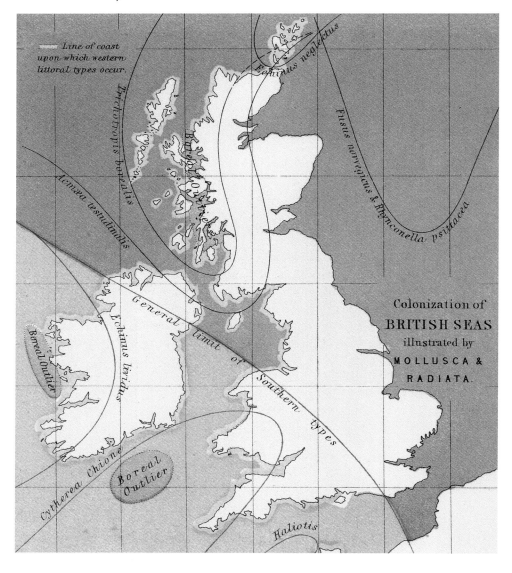

Biogeographical characteristics of the coast of the British Isles, including the range limits of some species. Drawn from research undertaken by Edward Forbes and fellow 'dredgers'. The map was published in A. K. Johnston's *The Physical Atlas of Natural Phenomena*. From the David Rumsey Map Collection.

Dredging studies continued throughout the latter part of the 19th century and, with the founding of the Liverpool Marine Biology Committee in 1885, much work was done in the area between the Isle of Man, Liverpool and Anglesey. Writing in 1887, William Herdman noted, 'It is hoped that a mass of observations will be accumulated which may be of use in establishing the geographical distribution of various forms, the nature of conditions which influence species and the relations existing between the different plants and animals'. Between 1885 and 1887, over a thousand species were recorded in the area.

Charles Kingsley (1819–1875).

Whatever successes the dredging campaigns achieved, British naturalists were frustrated in their exploration of the shallow seas. Charles Kingsley, in his volume *Glaucus or the Wonders of the Shore*, published in 1890, summed up a longing that many Victorian marine naturalists must have had when taking samples from the seabed:

> …. the riches of which have to be seen, alas! rather by the imagination than by the eye; for such spoonful's of the treasure as the dredge brings up to us, come too often rolled and battered, torn from their sites and contracted by fear, mere hints of what the populous reality below is like. Often, standing on the shore at low tide, one has longed to walk on and in under the waves … and see it all but for a moment.

SYSTEMATIC STUDIES – REMOTE SAMPLING

Towards the end of the 19th century, 'naturalists' had already made many advances in our understanding of seabed biology but many questions were being asked about the sustainability of fisheries, and there was an eagerness to understand what seabed and other marine life supported those fisheries. A centre for marine biological research was recommended in 1870 by Anton Dohrn and the British Association responded by appointing a committee for the foundation of zoological stations in different parts of the world. Marine biology became a more professionally established science and the opportunity to build a station dedicated to its study arose after the International Fisheries Exhibition in London in 1883. The following year, a meeting held at the Royal Society resulted in the founding of the Marine Biological Association of the United Kingdom and the opening of its laboratory at Plymouth in 1888. However, that institute was pre-dated in 1884 by the establishment of a marine station at St Andrews and a floating laboratory in a flooded quarry near Edinburgh. A year later, the latter was towed through the Forth and Clyde canal to Millport, where it became a precursor of the laboratory there which, in 1914, became the Laboratory of the Scottish Marine Biological Association. Meanwhile, in 1887, the Liverpool Marine Biology Committee established a marine station on Puffin Island off Anglesey. This continued operations only until 1891, when the centre of interest for marine research

Below left: the Marine Biological Association Laboratory in Plymouth in about 1900. The facade is very similar today. Image: from the Marine Biological Association archives.

Below right: the Scottish Marine Biological Association Laboratory at Millport. Date unknown but pre-1940 and before the east wing was built. Image: Scottish Marine Biological Association / Scottish Association for Marine Science.

in the area changed to Port Erin, Isle of Man. The marine laboratory established there by William Herdman in 1892 became a part of Liverpool University after World War I.

The marine laboratories established around Britain at the end of the 19th century would go on to undertake some very detailed studies identifying distinctive assemblages of species on and within the seabed. Foremost amongst those early studies was the work of Edgar Johnson Allen, Director of the Marine Biological Association at Plymouth. His maps and the accompanying lists of characterising species, published in 1899, are outstanding and have, in many subsequent studies, provided a starting point for identifying change. The work undertaken at Plymouth led to the compilation of the *Plymouth Marine Invertebrate Fauna* in 1904: the precursor of many such detailed accounts of local marine faunas.

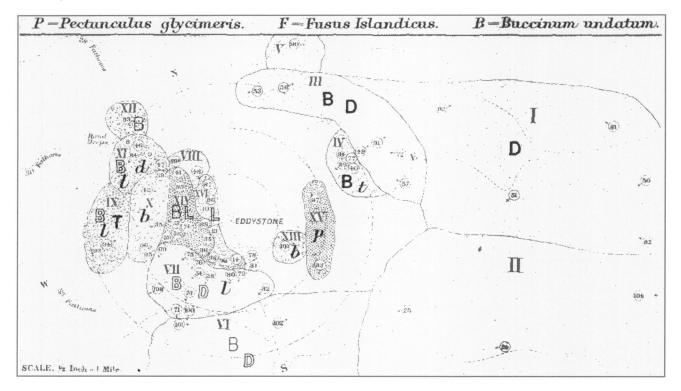

Work undertaken in 1913 off Denmark by C. G. Johannes Petersen would greatly influence the way in which surveys were undertaken around Britain. Petersen used a grab to provide quantitative samples of sediments rather than the 'superficial dredge' (the use of which he subjected to some scorn). This quantitative approach helped reveal different communities that were characterised by the conspicuous organisms, particularly molluscs and echinoderms, present in samples. R. Douglas Laurie and E. Emrys Watkin from Aberystwyth University College travelled to Denmark to study the methods being used by Petersen and came back with a Petersen grab. They were probably the first people to use the Petersen grab in British waters and described an area known as 'The Gutter' in Cardigan Bay in 1922, illustrating the communities present at two of the sample sites in the style of Petersen's drawings.

Part of one of the charts prepared by E. J. Allen showing the distribution of bottom types (habitats) and abundance of species in the region of the Eddystone reef. The Roman numerals refer to the bottom type. The letter codes are for the species named in red along the top (not all are shown in this clip). Solid large letters are for 'very abundant presence', hollow letters for 'only one or two stray specimens'. From the *Journal of the Marine Biological Association of the UK*.

A Petersen grab, believed to be the one used by Laurie and Watkin, being used to resample 'The Gutter' 70 years after the original survey in 1921. Image: Ivor Rees.

An illustration, from the paper by Laurie and Watkin published in 1922, of the most abundant species in one-tenth of a square metre sampled using a Petersen grab in Cardigan Bay in 1921. The community was described as the 'Amphiura region of Turitella-Amphiura community'. The species illustrated are the brittlestar Amphiura filiformis, the synaptid sea cucumbers Synapta (now Oestergrenia) digitata and S. (now Leptosynapta) inhaerens, the burrowing sea urchin Echinocardium cordatum, the polychaetes Pectinaria (now Amphictene) auricoma, Nephtys caeca, Sthenelais limicola and Evarne (now Harmothoe) impar, the siphunculid Phascolosoma procerum, and the molluscs Turritella communis, Kurtiella bidentata, Syndosmya (now Abra) alba and Corbula gibba. Image from Aberystwyth Studies.

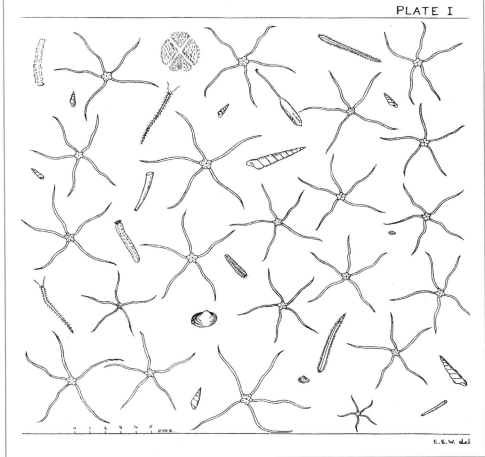

PLATE I

E.E.W. del

The work of Ebenezer Ford offshore of Plymouth Sound in the early 1920s, and of Frederick M. Davis and Alexander C. Stephen in the 1920s and 1930s in the North Sea, further followed the pioneering methods developed by Petersen.

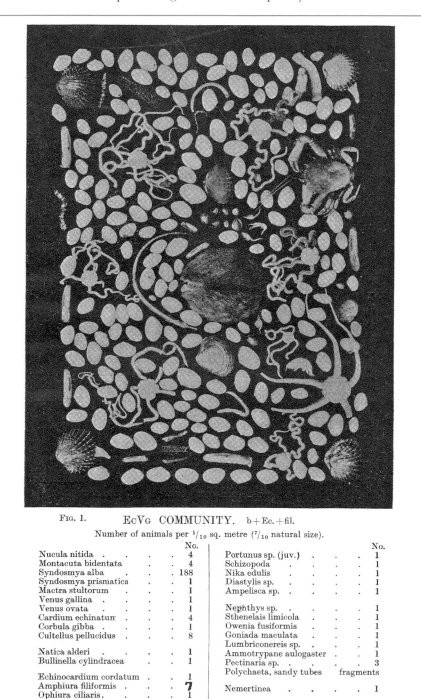

FIG. 1.   EcVg COMMUNITY.   b+Ec.+fil.

Number of animals per ¹/₁₀ sq. metre (⁷/₁₀ natural size).

	No.		No.
Nucula nitida	4	Portunus sp. (juv.)	1
Montacuta bidentata	4	Schizopoda	1
Syndosmya alba	188	Nika edulis	1
Syndosmya prismatica	1	Diastylis sp.	1
Mactra stultorum	1	Ampelisca sp.	1
Venus gallina	1		
Venus ovata	1	Nephthys sp.	1
Cardium echinatum	4	Sthenelais limicola	1
Corbula gibba	1	Owenia fusiformis	1
Cultellus pellucidus	8	Goniada maculata	1
		Lumbriconereis sp.	1
Natica alderi	1	Ammotrypane aulogaster	1
Bullinella cylindracea	1	Pectinaria sp.	3
		Polychaeta, sandy tubes	fragments
Echinocardium cordatum	1		
Amphiura filiformis	**7**	Nemertinea	1
Ophiura ciliaris	1		
Corystes cassivelaunus	1	Syngnathus sp. (juv.)	1

Station 63. Bigbury Bay { Borough Island, N.E. by E. } October 31st, 1922. Silty
                          { Bolt Tail, S.E. ½ S.   }   sand.

The contents of one grab sample collected by E. Ford (acknowledged as 'Naturalist at the Plymouth Laboratory' in his 1923 paper published in the *Journal of the Marine Biological Association*). The assemblage represented was named the '*Echinocardium cordatum - Venus gallina*' community, although the numerically most abundant species is the bivalve *Abra alba* accompanied by another common species in sediments, the brittlestar *Amphiura filiformis*. Image from the *Journal of the Marine Biological Association of the UK*.

There was extensive dredge sampling in the Firth of Clyde, especially in the early part of the 20th century and, later, around the middle of the 20th century, by scientists based at the marine station at Millport who published species lists in *The fauna of the Clyde Sea area* series. The seabed of lochs and the open coast of the highlands and islands of Scotland remained largely unexplored until the 1970s except, notably, by Edith Nicol for shallow lagoon habitats during the 1930s and some early grab sampling in Loch Linnhe in 1963 by Tom Pearson.

Dredge, trawl and grab sampling surveys adjacent to the Isle of Man, undertaken by Norman Jones in the late 1940s and the 1950s, would greatly improve knowledge of sediment habitats in the Irish Sea in general. The results from those surveys contributed to a classification of 'Marine Bottom Communities' named according to depth and to the bottom type in or on which they occurred. Later, in the mid-1950s, Alasdair McIntyre sampled the bottom fauna of fishing grounds off the east coast of Scotland, matching the results to Jones's classification.

It was not until marine biologists started to use diving in the 1930s and, later, towed camera sledges and then scuba equipment from the 1950s, that Kingsley's longing to 'see it all but for a moment' was to be fulfilled.

DIVING AND VIDEO SURVEYS

It was going to be cold (so a tin of treacle was consumed for energy), it was going to be wet (there was no relevant protective clothing, so woollen trousers and jumpers would have to do), but at least there was a telephone for dictating your scientific observations to a scribe at the surface (although that was needed to ask for 'more air, more air'). It was 1931 and a hardy group of young scientists based at the Marine Biological Association Laboratory in Plymouth were embarking on the first habitat survey by diving in Britain. Led by Jack Kitching and equipped with a specially constructed helmet that was fed air from a hand-operated pump, the three co-workers (the others were H. Cary Gilson and Thomas Macan), sampled a shallow gulley near Wembury at Tomb Rock in south Devon. That gulley is now known as 'Kitching's Gulley'.

Left: entering the water at Tomb Rock in Wembury Bay in 1931. Image: Jack Kitching.

Right: the south-east wall photographed in July 1986 during a survey commissioned by the Nature Conservancy Council. The conspicuous species were much as described in 1931, including calcareous sponges, Jewel Anemones *Corynactis viridis*, keeled tube worms *Pomatoceros triqueter*, sea squirts *Distomus variolosus*, and *Diplosoma spongiforme*. However, erect branching bryozoans (that dominated some surfaces in 1986) were very sparse in 1931/1932 and sea urchins *Echinus esculentus* were not mentioned in the 1934 paper.

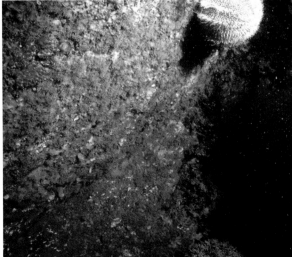

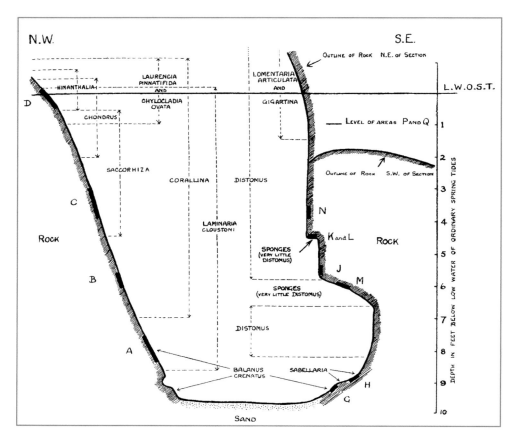

The drawing of the dominant species occupying the gulley from the 1934 paper (from the *Journal of the Marine Biological Association of the UK*).

I had the privilege to work with Jack Kitching at Lough Hyne in the late 1960s and early 1970s. I had just started my PhD studies and the sheltered underwater cliffs of the lough and the Whirlpool cliff (the name gives its character away), together with the adjacent wave-exposed open coast, were ideal for comparing the marine communities on bedrock that develop in different conditions of wave and tidal stream exposure. The drysuit and other equipment that Jack had progressed to after the war were still in storage in the wooden building that was one of the labs there. For me (and one looks back rather than realising at the time), the most important part of being one of the Lough Hyne team with Kitching in the lead was to understand the importance of teamwork, accuracy and good record keeping in research.

Almost 40 years earlier, Kitching had gone on to study kelp forests in the Sound of Jura and was especially interested in the reduction in light intensity and character with increasing depth and under the kelp canopy. That work was undertaken in September 1932 and in August 1933–1936. He dived to depths in excess of 15m and concluded that the distribution of species with depth was related to the strength of light and the attenuation of wave action.

Further studies using diving were to await the end of the war and would take advantage of some of the developments in diving equipment by the British and other navies. For Bob Forster, a staff member at the Marine Biological Association in Plymouth, that initially meant training in the use of an oxygen rebreather (his 'certificate of competence' dated 8th June 1951 reads 'This is to certify that

Bob Forster in the early 1970s. Image from the Marine Biological Association archives.

Mr G. R. Forster has completed a course satisfactorily and is competent in the use of the frog suit and the Self Contained Breathing Apparatus. He is capable of diving with the above apparatus in water up to a maximum depth of 33 feet.'). Bob published several descriptions of the characteristic macrofauna of hard substrata (rocks and the wreck of the *James Eagan Layne*). Fifty years on from his description of Hilsea Point Rock (then known as Stoke Point Rocks), I resurveyed the site and found it, with a very few significant differences, much as described in 1953.

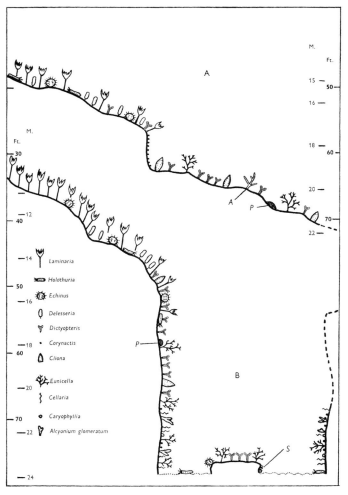

Diagrammatic profiles at Hilsea Point Rock in 1953 made by Bob Forster in one of the earliest studies of rocky subtidal habitats by scuba diving. Image from the *Journal of the Marine Biological Association of the UK*.

While Bob Forster was starting to use scuba diving for marine biology surveys, Norman Holme was using a dredge extensively in the English Channel to investigate the distribution of species. By the early 1960s, Holme was also taking advantage of new video technology together with colour photography and the advent of strobe flashguns to survey areas of level seabed. Those studies, using a towed sledge, gave a visualisation of the seabed and its associated marine life which is preserved in the archives at the Marine Biological Association. Extensive areas along the south coast of England were surveyed, the information providing a basis for later studies that might document change. We audited the image collection in 2004 and, although hoping to find photographs from the controversial spoil disposal site off Rame Head, we found

One of the towed photographic sledges designed by Norman Holme to survey the seabed in the English Channel and an image of the seabed at 78m depth south of the Eddystone Rock. Images: Norman Holme/Marine Biological Association.

none – the area had been used to dump munitions and there was most likely a fear of setting off an explosion that would be the end of the towed camera array at least.

By the mid-1960s, diving was becoming more than a tool for descriptive ecology and started to be used to help researchers understand how interactions between species influenced the communities that developed at a particular location. The studies on kelp biology undertaken by Joanna Kain (Jones) at the Marine Laboratory in Port Erin and from Millport were outstanding. They took place throughout the year and in wetsuits. Other marine biologists were beginning to use scuba. The description of seabed marine life along underwater transects by Wyn Knight-Jones and colleagues in the Menai Strait and at Bardsey laid the foundation for many later studies. Trevor Norton, after learning to dive at the Port Erin Marine Laboratory, moved to the University of Glasgow and dived around Argyll, describing the algal assemblages he found and identifying the characteristic algae of different depths and substrata. During the late 1960s, The University of London Sub-Aqua Club was assisting the compilation of the *Marine Fauna of the Isles of Scilly* by collecting samples and, in some cases, members were writing the separate accounts for major groups of organisms. Similarly, at Lundy, divers from the (then) North London Polytechnic were collecting algae to contribute to *A Survey of the Marine Algae of Lundy* – helping to record more species than at any similar-sized area in the British Isles.

A club for diving scientists was needed and the Underwater Association for Scientific Research (initially the Underwater Association of Malta, where the work of the founding members started) was formed in 1966. The Association brought together marine biologists, oceanographers and archaeologists: all using scuba techniques and sharing their knowledge and ideas. The Association was a key platform for what we now call 'networking' and was a source of inspiration to many of those who became leaders in the description of seabed habitats and of the species that inhabit them.

Gathering knowledge and creating information – the past 50 years or so

The exploration and description of seabed habitats has continued to a point where we can now organise data and create the information needed to inform environmental protection and management. Visualising those habitats has become much easier, and many people enrich their lives by scuba diving, or just by viewing some of the wondrous creatures and underwater landscapes around our shores through television programmes, magazines and books. There are still major gaps in our knowledge and much work remains to be done. This chapter explains how knowledge has been obtained in the past 50 or so years and how it has been organised to create information. It cannot adequately describe the range of significant studies, although much of the information that we have was brought together as a part of the Marine Nature Conservation Review and included in the volume *Benthic Marine Ecosystems of Great Britain and the North-East Atlantic*. A later chapter in this book 'Technology takes off' describes exciting new survey techniques and illustrates new discoveries.

REMOTE SAMPLING AND OBSERVATION

Deploying a Hamon grab from the Fisheries Research Vessel *Cefas Endeavour*.

Traditional techniques of trawling, dredging and grab sampling have provided, and will continue to provide, a great deal of our knowledge about the shallow seas around Britain. Those techniques were joined by seabed imaging via towed sledges developed in the early 1960s and, later, by remotely operated vehicles (ROVs). There has been extensive use of scuba diving in shallow waters from the 1960s onwards. The next section is dedicated to the rise of scuba as a survey tool.

Notable programmes of survey and sampling have contributed to understanding what is where, identifying and refining our knowledge of seabed habitats and their associated assemblages of species. Information has also come from fisheries laboratories, which were documenting the character of fishing grounds and following any changes in them. Other studies have come from the oil and gas, renewable energy, aggregates and other industries, charged with assessing likely impacts and monitoring actual impacts on the seabed. The most valuable have been dedicated surveys aimed at characterising the biology of areas, to improve our scientific understanding of what is where, and especially to provide information that will underpin marine conservation.

In Shetland, the Institute of Terrestrial Ecology grab sampling surveys in 1974, undertaken by the Scottish Marine Biological Association, extended throughout the voes although later surveys, commissioned by the oil industry, were mainly in Sullom Voe. By 1981, when Tom Pearson and Tasso Eleftheriou published their account of the benthic ecology of Sullom Voe, 334 species of macrofauna had been identified from grab samples. Monitoring continues in Sullom Voe and, by 2014, the species count for sublittoral sediments Sullom Voe had reached about 1,270. Other oil industry surveys inshore were undertaken in Orkney, in the Moray Firth, in Southampton Water and in Milford Haven especially: all adding to our knowledge of shallow sea sediment communities.

The Firth of Forth and the Forth Estuary were extensively sampled in the 1970s and 1980s, in particular by Mike Elliott and Paul Kingston, revealing four major community types in sediments in the Firth of Forth and zones reflecting salinity changes in the estuary. Offshore in the North Sea, there were extensive grab sampling surveys being undertaken and published, especially in the 1980s. However, surveys undertaken for monitoring purposes around North Sea oil and gas installations are rarely published. Further south, extensive sediment communities in the shallow sublittoral zone were being sampled, mainly in relation to aggregate industry activities, including various studies to assess recovery of seabed communities from sand and gravel extraction. The Wash was an important area of conservation but little was known of the character of its subtidal assemblages until sampling was undertaken by Frances Dipper, Sarah Fowler and Robert Irving for the Nature Conservancy Council (NCC, responsible for designating and managing National Nature Reserves and other nature conservation areas in Britain between 1973 and 1991). The marine inlets of Essex have been described for their subtidal biota and various communities. Off the south coast, the aggregate industry is active and significant studies have accurately mapped sediment communities based on grab sampling and imaging. Further west, Southampton Water has been extensively sampled. Whilst the towed sledge images collected by Norman Holme in the western English Channel may provide valuable data for mapping of seabed habitats and communities, that outcome has not yet been achieved.

Sampling underway in the Irish Sea from the RV *Prince Madog*. The image shows sieving teams and grab preparation. Image: National Museum Wales.

A beam trawl being recovered.

A typical trawl sample.

The BIOMÔR programme of the National Museum Wales extensively sampled areas of the southern Irish Sea, Bristol Channel and Severn Estuary over the period 1985 to 2010. A variety of equipment was used to collect samples from sediment habitats including grabs, dredges and trawls. Towed video sledges were also used to obtain images, some of which are shown in the 'Sediments' section of the 'Habitats' chapter. Images: National Museum Wales.

A grab sample being recovered.

Emptying the sample collected by a dredge.

The surveys undertaken by the Natural Environment Research Council (NERC) in the Bristol Channel and Severn Estuary mapped seabed assemblages named according to the communities of Petersen. Similarly, surveys from the then Marine Science Laboratories at Menai Bridge undertaken by Ivor Rees and colleagues identified Petersen communities in the Irish Sea. Very extensive surveys in the southern Irish Sea and outer Bristol Channel were completed as a part of the BIOMÔR (National Museum of Wales) programme. Those surveys mapped the sediment types and associated communities over large areas.

Sediment types in the southern Irish Sea. One of the maps generated as a part of the BIOMÔR programme and based on in situ sampling. The map appeared as Fig. 5.4 in BIOMÔR 2 (The Southwest Irish Sea Survey). Reproduced with permission of the National Museum Wales.

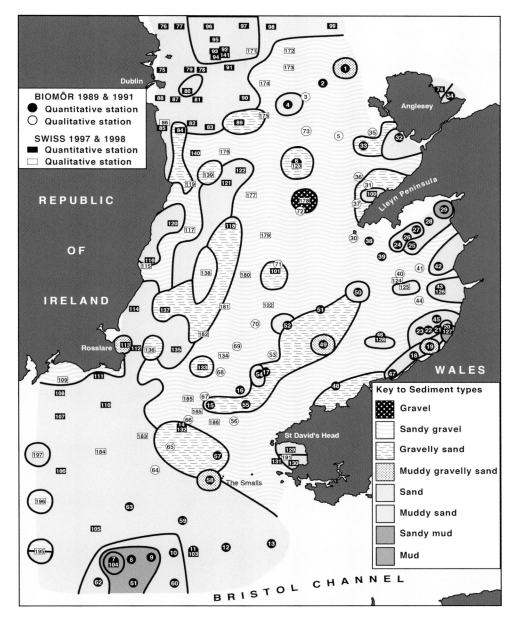

The eastern basin of the Irish Sea including Liverpool Bay received little attention after the early dredging studies in the late 19th and early 20th century until there was an upsurge in activity in the 1970s. This activity was directed mainly into studies of the environmental effects of the dumping of sewage sludge and dredge spoil, and the discharges of polluting effluents. Similarly, it was concern about the potential effects of industrial effluents and sewage disposal that led to surveys off Cumbria, in the Solway Firth and in the Clyde.

Sea lochs seemed largely ignored by biologists studying subtidal seabed communities until the surveys commissioned by the NCC and its successors from the 1970s onwards. However, notable work was carried out from the Scottish Association for Marine Science in Loch Etive and Loch Creran near Oban, where

John Gage and colleagues extensively sampled and described the fauna, mainly by taking core samples during dives. The sediment communities of Loch Linnhe and Loch Eil adjacent to Fort William were intensively sampled through the 1970s and 1980s by Tom Pearson and colleagues. Their sampling was targeted in particular at understanding the impact of a pulp mill's effluent on seabed marine life. The results revealed what is now a classic illustration of the gradient of change in relation to increasing distance from the source of organic enrichment: a gradient that, today, is repeated adjacent to many fish farms. There were grab sample surveys of sea lochs by the fisheries laboratory in Aberdeen in the 1960s but the laboratory notebooks listing the fauna have not been seen in recent years.

The most recent and ongoing large-scale surveys, part of the Marine Ecosystems Research Programme (MERP), aim to fill gaps in knowledge so that directives, conventions and statutes are better informed. The programme, which looks especially at ecosystem processes, is jointly funded by the UK's NERC and Department for Environment, Food and Rural Affairs (Defra).

A Remote Operated Vehicle (ROV) being deployed by Scottish Natural Heritage off Cumbrae in the Firth of Clyde to survey seabed habitats and associated species. ROVs are used extensively to visualise seabed habitats and especially in monitoring structural integrity and fouling on offshores structures. Image: Scottish Natural Heritage.

Swath acoustic data, such as this bathymetric map produced from a multibeam echosounder, may be used to identify biogenic features where those features produce a distinctive signal. Here, the Horse Mussel *Modiolus modiolus* reef is visible diagonally across the sonar image obtained off the Lleyn Peninsula in north Wales. Image: HABMAP Project.

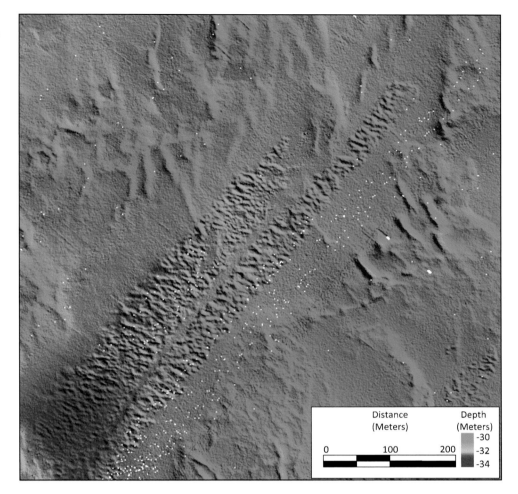

Distance (Meters)			Depth (Meters)
0	100	200	-30
			-32
			-34

While grab and dredge sampling reveal detailed information on the biology of spot areas of sediments, and photographs reveal surface biology where underwater visibility is favourable, sound is virtually the only useful imaging technique that penetrates any distance in seawater. A revolution that has occurred in seabed mapping has been the development of swath acoustic techniques achieved through multibeam sonar systems that scan wide areas of seabed. The collection of multibeam echosounder data continues to increase and, in 2014, that data was available for about 200,000km^2 or about 26% of the UK Exclusive Economic Zone. It immediately gives the user information on depth and reveals features such as sand ripples, cliffs, canyons and wrecks that may be worth investigating to see what life exists there. In some cases, the signal may be sufficiently distinctive that, with appropriate ground truthing, it can indicate the presence of certain habitat types. A single beam system known as RoxAnn was effective in identifying some biogenic reef types such as the Horse Mussel *Modiolus modiolus* beds off the Lleyn Peninsula. Put simply: because the mussels are rounded or end-on to the sound wave, they return a weaker echo than the hard lag cobbles and gravel surface around them. They also accumulate faecal mud amongst the mass. Because the surface is rough and complex, there is attenuation of the first return, which gives a measure of roughness. Plotting roughness against hardness and then mapping the resulting classification category allowed the extent of the *Modiolus* bed along survey lines to be indicated. However, converting acoustic data into maps of seabed substrata is rarely an easy task. Leading work is being done by the Centre for Environment, Fisheries and Aquaculture Science (Cefas) and, in their 2015 paper in the *Journal of Sea Research*, Markus Diesing and David Stephens explain how ground-truthed sample data from video and grab samples can be used to train machines to interpret swath acoustic data. Novel techniques such as these can be very helpful, as they interpret vast amounts of data automatically in a consistent manner, reducing human interpretation judgement errors. Acoustic technology, data resolution and interpretation methods are constantly improving, but predicting the distribution of biotopes from acoustic data remains a challenge.

There are now many maps of seabed habitats around Britain but few that are from ground truthed data (and therefore likely to be accurate). The user of subtidal seabed habitat maps needs to be cautious – many have been produced by extrapolation from sparse ground truth data and many more from physical data with the aid of algorithms: confidence in their accuracy is often low.

SCUBA DIVING TAKES-OFF

The desire of Victorian naturalists to 'see it all but for a moment' was to be fulfilled nearly a hundred years after that hope was expressed. The first significant scientific project involving recreational divers was Operation Kelp in 1967, coordinated by David Bellamy and collecting data from 27 locations around Britain. Many others followed, with divers often collecting samples of plants and animals for specialists to identify. Many of those amateur naturalists went on to become highly respected specialists in their field.

By the early 1970s, scuba diving was becoming a widely used tool in marine ecological research and (perhaps especially for new arrivals like me), an opportunity to explore, describe and experiment in habitats and with species that had previously been inaccessible. An open door beckoned, and with my passions for marine natural history, diving and photography, I was well-equipped to go through it. There was always luck involved in doing what I wanted to do and my studies, while an undergraduate, on the ecology of the Devonshire Cup Coral *Caryophyllia smithii*, helped to convince Professor Denis Crisp that I would be able to tackle a PhD. That PhD became (after a failed attempt to pursue studies of reproduction and settlement biology of hydroids) a study of *The Effects of Water Movement on the Ecology of Sublittoral Rocky Areas*. I was greatly helped in pursuit of my studies around Anglesey and Lundy and at Abereiddy Quarry in Pembrokeshire by Richard Hoare and Dave Lane but also by a high degree of tolerance of my 'determined' approach to diving by relevant staff at the then University College of North Wales Marine Science Laboratory at Menai Bridge.

The 1970s and early 1980s were the heyday of using scuba for undertaking scientific studies. Through the newsletters, journal and annual meetings of the Underwater Association for Scientific Research, we were alerted to new developments and found out who was doing what. Never underestimate the value of networks, friendships and what some consider junkets – contributing to conferences – in developing your field of study.

Early research diving. The author at Bessy's Cove on The Lizard in 1969.

The diving equipment used for research was (and still is) essentially the same as that developed for recreational diving. In the 1950s and 1960s, that meant thin neoprene wetsuits (with yellow tape on the seams), a surface support jacket, a single air cylinder and a regulator, most likely without a contents gauge – well, you would know when air pressure was getting low as you started to 'suck on your cylinder'. By the 1970s, wetsuits were getting thicker and we had adjustable buoyancy jackets. Diver-ecologists now stay dry and warm in drysuits, have computers to calculate dive duration and decompression, and have oxygen-enriched air (Nitrox) or rebreathers to extend dive times. Health and safety legislation has increased the bureaucracy of diving and sometimes required personnel and equipment that are additional to what may be considered necessary – but the days of having only one source of air and no contents gauge are over. Photographic equipment has also improved enormously. I started with an Exa single lens reflex camera in a (slightly leaky) Perspex housing made in the workshops at Westfield College. Nikonos film cameras with bulb flashguns followed, then with electronic flashguns, and, since about 2000, digital cameras. The great majority of underwater images in this volume are taken with a digital SLR in a housing with twin electronic flashguns.

Newsletters and reports (in the series *Progress in Underwater Science*) from the Underwater Association included articles on survey methods for biological surveys (amongst much else). Those methods were developed after some trial and error and some hair-raising expeditions. The survey of Shetland in 1974 by Bob Earll and Chris Lumb involved shore dives at randomly selected examples of particular coastal types and at Ordnance Survey grid intercepts of the coast – so, wherever those intercepts were across bogs or down steep cliff faces, there they went. The stratified random sampling methodology was derived from terrestrial approaches to surveying and was aimed at providing results that satisfied statistical test requirements. The methodology didn't catch on and most surveys employed a more pragmatic approach of simply trying to ensure that, based on inspections of maps and charts, the range of seabed habitats in a designated area (for instance, an island archipelago, a length of physiographically similar coastline) would be surveyed and, for practical reasons, to depths generally shallower than 40m. Survey divers were also briefed to record what they saw in distinctly different habitat types, rather than at specific depths or lengths along a transect. Only conspicuous species that could be recognised underwater were generally recorded, although samples were needed of plants and animals that could not be identified in situ.

In the early 1970s at least, there seemed to be reluctance by research funding bodies to support surveys using scuba diving – perhaps it looked too much like having fun. Nevertheless, diving was being used to sample underwater marine life for inventories of species and to support work towards the designation of marine nature reserves at Lundy and Skomer and a few other locations. In 1977, there were a couple of turning points. The NCC had discovered marine nature and appointed Roger Mitchell to join their Chief Scientists Team and commission, from 1977 onwards, a series of research programmes that would inform marine conservation. Underwater Conservation Year (UCY) also took place in 1977 and divers were encouraged by the project coordinator, Bob Earll, to do a spot of what was described as 'underwater birdwatching'. There were projects and there were expeditions. Record cards were designed and completed and there were specialist studies on the abundance of sea urchins and the creation of photographic guides to identify marine life. Very importantly, there was feedback and acknowledgement of the value of amateurs. Underwater Conservation Year transformed into the Underwater Conservation Society with its headquarters in a shed at the bottom of Bob Earll's garden and then onward to proper offices in Ross-on-Wye, where it evolved into the highly successful and influential Marine Conservation Society (MCS).

Many of the research techniques that were developed in the 1970s and early 1980s are essentially the ones that we use today, including surveys of conspicuous species using semi-quantitative abundance scales, taking photographs (although, now, no longer restricted to 36 frames on a film cassette) and writing our notes on writing slates made of plastic laminate. And, on the way, a few things that didn't work – underwater tape recorders often flooded and the description of what was being seen was scarcely understandable. Some methods we used a few times but not much more – being towed along with a planing board over the seabed was good fun but didn't catch on for extensive studies, diver-operated underwater suction samplers worked really well but the samples were so rich in species that there was rarely time to sort and identify the sampled material. Some techniques have been adjusted and work well in a different form. Fixed photographic quadrats to monitor change started to be used in the early 1980s, although using two cameras to obtain stereophotographic images had little to recommend it.

Left: a Barnett and Hardy air lift suction sampler for sediments wielded by Jim Wilson at Lundy in 1977.

Right top: a Hiscock and Hoare air lift suction sampler for rock epibiota.

Right bottom: a planing board used by a diver being towed behind a boat to survey extensive areas, pictured at Lundy.

Left top: Iain Dixon core sampling along a pollution gradient away from a fish farm in West Voe, Shetland. Diving enabled accurate location of samples along a transect.

Left bottom: photography was extensively used, both to illustrate what was being seen and to re-survey fixed sites – Robert Irving in the Isles of Scilly in 1985.

Right: rigged to record but, all too often, the tape recorder housing flooded and the dictation was anyway incomprehensible.

There were significant survey programmes in the 1970s and 1980s, predominantly using scuba diving, commissioned or supported by the NCC. Surveys around St Abbs involving Bob Earll and Sue Hiscock (now Scott), and others in Northumberland, especially work undertaken by Bob and Judy Foster-Smith. In Sussex and Kent, the plans for a Channel tunnel gave rise to subtidal surveys undertaken by Liz Wood and others. Also in Sussex, Robert Irving was engaging amateur divers in surveys that described the underwater habitats there. Dorset County Council, the NCC and Dorset Wildlife Trust commissioned the Dorset Underwater Survey, which ran from 1976 to 1978. Its diving studies identified 'epilithic associations' and involved especially Ken Collins, Clive Roberts and Iain Dixon. Moving further west, there were two major surveys, mainly undertaken by diving and overseen by me: the South-west Britain Sublittoral Survey and the Harbours, Rias and Estuaries Survey. We had a variable feast of surveyors, but particularly worthy of mention are Iain Dixon, Sue Hiscock (now Scott), Annette Little, Christine Maggs, Jon Moore and Dale Rostron. Those surveys relied on facilities provided by the Field Studies Council at Orielton Field Centre in Pembrokeshire. In Wales, there were surveys around Skomer, especially involving Blaise Bullimore, and of Sarn Badrig. Throughout those surveys, distinctive assemblages of species were being described but we were still a long way off developing a classification of those assemblages.

Descriptive surveys of the marine life in the shallow seas around Britain took off with the inauguration of a Marine Nature Conservation Review of Great Britain (MNCR) in 1987 by the NCC. I was delighted to be appointed as head of the team

Quantitative sampling from hard substrata

During the 1970s and 1980s, quantitative samples from subtidal rocky areas were being successfully collected using an airlift suction device but there were only a few projects, especially at Lundy. An experimental study to determine minimum sampling areas to obtain mean densities of organisms within statistical limits by taking and analysing samples on the close-to-homogeneous surface on the port side of the wreck of the *MV Robert* at Lundy revealed that a practically impossible number of samples would be needed to obtain those mean densities for most organisms. My patience doesn't extend to picking separate organisms out of a pile of stuff but Dale Rostron did an excellent job. The collection of quantitative samples from subtidal rocky areas never took off, most likely because of the tedious and time-consuming nature of analysing them.

A portable airlift suction sampler in action on the deck of ex-HMS *Scylla*. Air rising in the vertical pipe creates suction which pulls water and the sample into a mesh bag in the sample chamber. What a sample looks like: 0.1m² of growth collected from 15m depth on the side of the *MV Robert* off Lundy. Fourteen samples were analysed and each sample held between 29 and 91 species with a total of 192 animal taxa recorded from the 1.4m² area sampled.

and, over the next 11 years, a programme of surveys, both by the team and through external contractors, identified what was where around our coasts, how to classify what we recorded, and how that information could be used to support marine conservation. Much of what I know about what is where on the shallow seabed around Britain comes from fieldwork as a part of the MNCR. The rationale and methods were published in 1996 and a volume summarising our knowledge of seabed (benthic) marine ecosystems in 1998. After fieldwork finished in 1998, team members went on to produce a series of *Regional Summaries*.

Answering an enquiry I made in 2004 about how many locations the MNCR had included in its surveys, the figures of 111 surveys, including around 3,000 associated locations and 8,800 sample stations, was determined. Although the MNCR 'finished'

The Marine Nature Conservation Review incorporated surveys from Muckle Flugga, the furthest north island in Britain, to Bishop Rock in the Isles of Scilly in the far south-west. Image of Muckle Flugga: David Connor/JNCC.

Describing the abundance of species at a survey station in 1988 during the MNCR survey of Shetland. The blue box is an early diving computer – the Orca Edge. Image: Sue Scott/JNCC.

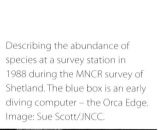

in 1998, it was never complete. Much survey work remains to be done, as there are gaps in knowledge that predictive algorithms cannot fill. Most importantly, all of the information collected through the MNCR was held in a database that could be interrogated and used to support environmental protection and management.

MNCR surveys also described sediment communities and, especially where sediments were localised amongst rocks, used a suction sampler. Here, a suction sampler is in use at Craignish in west Scotland. Image: Sue Scott/ JNCC.

Kelp holdfasts as sampling units

Another way of sampling to compare species richness and abundance between survey locations is to collect holdfasts of *Laminaria hyperborea* and to use the holdfasts as comparable sampling units. The technique, requiring scuba diving and a rapid transfer of the holdfast to a polythene bag, was used by David Scarratt working at the Marine Biological Association, and was further developed in a series of studies by Geoff Moore working mainly on the north-east coast of England. Those studies demonstrated the importance of water turbidity in determining the assemblages of species that developed in and on the holdfasts. Moore recorded 389 macrofauna and meiofauna species from all of the samples. Kelp holdfasts also proved a valuable sampling unit when investigating pollution gradients. Richard Hoare and I found that within about 55m of an acidified and halogenated effluent on the north coast of Anglesey, 30 macrofauna species were sampled from holdfasts, whereas at over 1km from the effluent, 70 species were found in holdfasts of comparable size. The dominant species present along the pollution gradient were also very different. The importance of kelp plants, including holdfasts, as sampling units was reviewed by Harry Teagle and others in the *Journal of Experimental Marine Biology and Ecology* in 2017. In the same study, *L. hyperborea* had, further, been collected from 12 locations spanning nine degrees of latitude from Orkney to Plymouth, to better understand biogeographic patterns and the influence of environmental factors (light, wave action, temperature) on community structure.

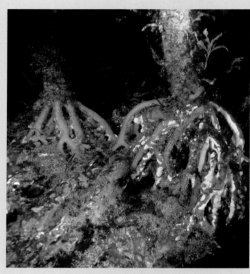

Holdfasts of *Laminaria hyperborea* photographed south of the breakwater in Plymouth Sound.

Perhaps nothing good lasts forever and, in the same year as the MNCR was 'finished' in 1998, the Underwater Association was wound-up. Many of its members were already contributing to the Porcupine Marine Natural History Society (named after HMS *Porcupine*, a hydrographic surveying vessel active in the mid-1800s) and Porcupine continues to provide opportunities for field work and for sharing knowledge of marine life and habitats. Many of the pictures in this book were taken on Porcupine field trips. The statutory nature conservation bodies in Britain continued to commission some fieldwork after the shut-down of the MNCR, but were much more concerned with the implementation of the various statutes, directives and conventions that the UK is signed-up to. That especially included surveys to report on whether 'Favourable Conservation Status' in Special Areas of Conservation was being maintained. The Seasearch project developed in the late 1980s by Bob Earll (then Head of Conservation at MCS) and Roger Mitchell (Head of Marine Conservation Branch in the NCC) would later help to pick up the traces of MNCR work and contribute enormously to the information base needed for the identification of potential protected areas. Seasearch surveys are coordinated nationally by MCS and organised by regional coordinators including the Wildlife Trusts. The programme had a major financial boost, particularly from the Heritage Lottery Fund, and a new project was officially launched in June 2003. In recent

With a growing need in the statutory nature conservation bodies to report on whether MPAs were in 'Favourable Condition', scuba played an essential role in monitoring. In the Skomer MPA, that monitoring has been underway since the 1970s and now provides detailed information of the changes that occur underwater. Here, Kate Lock and Phil Newman plan their underwater work: transects at marked locations are laid-out, and Phil takes photographs for later analysis.

years, coordinated by Chris Wood (who organised my early diving trips to Cornwall in the late 1960s), Seasearch has thrived and provides the opportunity for non-specialists and specialists to contribute valuable information to conservation action in their spare time – something that we now call 'Citizen Science'. Many of the pictures in this book were taken on Seasearch field trips.

Writing up the results of a Seasearch survey in 2008.

HOW WE MADE SENSE OF IT ALL (BIOTOPES)

A thirst for knowledge is a fundamental human trait, and we use that knowledge to understand how things are organised and how they work. We need to make an incredibly complicated world as understandable as possible, and that means developing classifications and codes that simplify in a logical way what might seem chaotic. Linnaeus did that for the naming of species in the 18th century. Botanists on land created classifications for plant communities during the 20th century which, for purposes of supporting nature conservation, became the *National Vegetation Classification* commissioned in 1977. Seabed marine life required a different approach, as many or most distinctive assemblages of species were characterised by animals. This section of the book describes how progress was made towards the classification of distinctive seabed biotopes.

During the latter half of the 19th century, marine naturalists were beginning to see how different assemblages of species could be found in different locations and in different habitats. In 1890, Charles Kingsley, writing in his book *Glaucus*, had recognised that 'the sea-bottom, also, has its zones, at different depths, and its peculiar forms in peculiar spots, affected by the currents and the nature of the ground'. Since that time, marine biologists have been identifying patterns in the occurrence of distinct assemblages of species and giving those assemblages names.

The studies undertaken between 1895 and 1898 by Edgar Johnson Allen from the Marine Biological Association Laboratory in Plymouth, using dredges and trawls from various grounds off the south Devon coast, recorded the abundance of a total of 360 species described, for the first time it seems, in distinctive assemblages. One of the charts that Allen produced is included in the previous chapter (page 14).

In 1915, C. G. Johannes Petersen produced a map showing the likely distribution of what were described as seabed 'communities' in the north-east Atlantic, including around Britain. Petersen recognised two principal and ecologically different groups of benthic marine animals: the *epifauna*, comprising all animals living upon or associated with rocks, stones, shells, piling and vegetation, and the *infauna*, comprising all animals inhabiting the sandy or muddy surface layers of the sea bottom. Petersen's

Making a biotope. The image on the left is bedrock at 20m depth in the submerged river gorge near the entrance to the Tamar estuary. The habitat is described as 'turbid tide-swept sheltered circalittoral rock'. On the right is the community of species that develops there and is described as 'cushion sponges, hydroids and ascidians'. The community of species plus the habitat on/in which it occurs creates the biotope name: 'cushion sponges, hydroids and ascidians on turbid tide-swept sheltered circalittoral rock'. Image widths *c.* 80cm. (The bare bedrock was created by mooring chains from a navigation buoy.)

'communities' were 'a regularly recurring combination of certain type animals, as a rule also strongly represented numerically' but he acknowledged that 'in the present state of knowledge, it is impossible to show how intimate the mutual relationships are between the animals of the sea in the different cases'. Thus, some workers preferred to refer to 'assemblages', suggesting no particular associations between the inhabitants: they thrive in the same physical and chemical (salinity) conditions and geographical location. Petersen's studies, undertaken mainly around Denmark, used a grab to provide quantitative samples of sediments to reveal different 'communities'. Later, sampling by, amongst others, R. Douglas Laurie and E. Emrys Watkin in Cardigan Bay, Ebenezer Ford offshore of Plymouth Sound and Frederick M. Davis and Alexander C. Stephen in the North Sea, followed the pioneering methods developed by Petersen to describe the sediment communities in British seas.

The quantitative studies of seabed sediment communities undertaken by Davis in the southern North Sea during the 1920s led to the suggestion that, in relation to the distribution of communities, 'the simple number of the soil groups will show what species may be expected therein', thus stating the primary importance of sediment type in determining the communities likely to present. The results of these early studies and of some later ones such as those of Norman Jones from south of the Isle of Man and published in 1950 were brought together by Gunnar Thorson in 1957 to describe the level-bottom animal communities and their distribution from throughout the world including especially those noted for the north-east Atlantic. His Memoir, published by the Geological Society of America, has many important points of guidance for anyone now working on the description of seabed communities. Petersen's communities continued to provide the catalogue of level seabed communities well into the 1970s; for instance, in studies in the Bristol Channel by Richard Warwick and John Davies published in 1977.

Reef habitats were largely being ignored by biologists studying seabed communities because grabs and dredges could not adequately sample them. By the 1950s, descriptions were beginning to be published of hard substratum species assemblages at specific locations: work carried out especially by Bob Forster near Plymouth and Wyn Knight-Jones in north Wales using scuba diving. My PhD studies in the early 1970s identified specific and recurrent assemblages that occurred in relation to exposure to wave action and tidal current strength. Later, the studies being commissioned by the NCC from the mid-1970s were not only listing the species present at specific subtidal locations, but were identifying what seemed to be recurring groups of species, i.e. assemblages or communities. The work was used extensively in a publication, *The description and classification of sublittoral epibenthic ecosystems*, written by myself and Roger Mitchell and published in 1980 in a Systematics Association Special Volume. That work outlined the approaches and standards including terminology that would be used in the Marine Nature Conservation Review of Great Britain (MNCR) which started in 1987.

One of the objectives of the MNCR was to develop a 'marine classification system to underpin interpretation of data, assessment of conservation importance and management of marine areas'. It was Mark Costello, then at Trinity College Dublin, who saw a route that would bring together European marine biologists to develop a classification of what had, by the early 1990s, become known as 'biotopes'. That EU-funded programme was BioMar and it was in November 1994 that 49 scientists met together in Cambridge to establish the principles behind the classification that now identifies separate seabed habitats and their associated species. The work was taken forward by David Connor and the MNCR team.

The MNCR had accumulated a great deal of survey data from its own work but, very importantly, had been adding survey data from a wide range of sources to a computer database (new technology at the start of the MNCR) that it had used since the beginning of the programme. BioMar encompassed the island of Ireland where much survey work was also underway. The data was processed to identify species groupings that commonly occurred together and then, very importantly, a sense test applied to the computer output by experienced seabed ecologists before a distinctive biotope was identified. Existing classifications (mainly from sediments) and classifications being used in France were taken into account. The first working version of the classification had been distributed in April 1994 and there were several subsequent iterations. The MNCR biotopes also became the basis for a classification for the European Seas within the European Nature Information System (EUNIS). The two classifications are fully compatible and both are used. Examples of parts of the hierarchical classification with EUNIS and MNCR codes are shown next.

A [Level 1]: Marine habitats

A4 [Level 2]: Circalittoral rock and other hard substrata ('CR' in the MNCR classification)

A4.1 [Level 3]: Atlantic and Mediterranean high energy circalittoral rock ('HCR' in the MNCR classification)

A4.13 [Level 4]: Mixed faunal turf communities on circalittoral rock ('Xfa' in the MNCR classification)

A4.131 [Level 5]: Bryozoan turf and erect sponges on tide-swept circalittoral rock ('ByErSp' in the MNCR classification)

A4.1311 [Level 6]: *Eunicella verrucosa* and *Pentapora foliacea* on wave-exposed circalittoral rock ('Eun' in the MNCR classification)

A circalittoral reef biotope distinguished by the most abundant or characterising species and by the physical and chemical (salinity) conditions in which it occurs. This is the biotope in the hierarchical classification demonstrated above: '*Eunicella verrucosa* and *Pentapora foliacea* on wave-exposed circalittoral rock' (A4.1311 /CR.HCR.XFa.ByErSp.Eun). Photographed at a depth of 22m below chart datum in mid-August 2007 off Stoke Point, east of Plymouth. Image width *c.* 1.2m.

A5 [Level 2]: Sublittoral sediment ('SS' in the MNCR classification)

   A5.4 [Level 3]: Sublittoral mixed sediments ('SMx' in the MNCR classification)

      A5.44 [Level 4]: Circalittoral mixed sediments ('CMx' in the MNCR classification)

         A5.441 [Level 5]: *Cerianthus lloydii* and other burrowing anemones in circalittoral muddy mixed sediment ('ClloMx' in the MNCR classification)

Sediment biotopes are characterised by grain size and mix as well as the species present in the sediment. Identification therefore usually requires sampling to identify infaunal species, although there are often clues in the conspicuous organisms that can be seen at the surface. Here, the biotope '*Cerianthus lloydii* and other burrowing anemones in circalittoral muddy mixed sediment' (A5.441 / SS.SMx.CMx.ClloMx). Photographed at a depth of 32m below chart datum at Inner Penninis Point in the Isles of Scilly.

There are about 280 different biotopes that are likely to be identified in subtidal seabed habitats in the shallow seas around Britain. They are from a much larger classification that includes about 500 entities in a six-layered hierarchy and includes the shore as well as shallow seas. That hierarchical approach allows surveys and maps to be made at the level of detail needed to inform decision-making (for instance, a broadscale map at a coarse level of the classification provides large coverage for marine spatial planning but a more detailed biotope map may be needed for protected site management). In this book, the codes are from the EUNIS Habitat Classification 2007 (revised descriptions 2012) and the Marine Habitat Classification for Britain and Ireland version 15.03 (the MNCR classification). The Joint Nature Conservation Committee (JNCC) is the agency currently responsible for the ongoing development and use of the Marine Habitat Classification for Britain and Ireland.

A particular biotope may occupy an area of many square kilometres in the case of offshore level sediments but, in inshore areas, will usually be identified in areas of a few (say 25) square metres and, where steep rock surfaces occur in bands only of one or two metres deep. They can also be distinguished in very small areas, such as under overhangs or on a boulder on sediment.

Perhaps I have made the recognition of distinctive biotopes seem easy – it often is not: there will be intermediates and there will be assemblages that do not fit the classification. The classification will evolve but it is a robust tool for use in describing what the seabed wildlife at a location is, for identifying how widespread or how rare a particular entity is, for comparing like with like in assessing the best or richest examples, and for linking to sensitivity assessments.

Shaping the seabed environment

Environmental conditions at any particular location are important in determining what lives there – they shape the habitat by creating conditions that favour the presence of particular species. Those conditions may be modified by human activities but this book is about what to expect in conditions that are as close as possible to natural.

The concept of 'habitat' was expressed very clearly by leading plant ecologist Arthur George Tansley:

> the term habitat may be taken to mean 'the sum of the effective conditions under which the plant or the community lives' …. Every species and every community has a certain range of habitat, which may be wide or narrow.

That definition could equally be applied to the seabed.

Foremost amongst the conditions that create a particular underwater habitat are seabed type (bedrock to mud), illumination (the amount and type of light reaching the seabed depending on depth, water turbidity and shading), the strength of wave action, the strength of tidal currents and the salinity of the water. Biological interactions are many and often poorly understood but include, most conspicuously, habitat provision, predation including grazing, and shading. From time to time, weather events may re-set conditions. Students and lecturers can turn to a large number of publications that provide technical explanations of environmental conditions but, here, broad descriptions of major factors will suffice to give context to the sections on specific habitat types.

My conclusions regarding the character and effects light were summarised in the chapter 'Aspects of the Ecology of Sublittoral Rocky Areas' in the volume *The Ecology of Rocky Coasts* edited by Geoff Moore and Ray Seed, published in 1985, whilst the effects of wave action and tidal currents were summarised in 'Water Movement' in the book *Sublittoral Ecology: the Ecology of the Shallow Sublittoral Benthos* edited by Bob Earll and David Erwin published in 1983.

BRITAIN'S PLACE IN THE WORLD

The geographical distribution of species is determined by the major oceans within which they are confined and by more local conditions of temperatures that they are attuned to thrive or survive in. There may be other factors such as the availability of suitable food and there may be barriers to spread of adults and larvae, such as extensive areas of sediment between rocky seabeds that mean, within the lifespan of a larva, they will not reach a suitable substratum to settle in a different region or sub-region. The description of different areas distinguished by the plants and animals that live there is described as 'biogeography'.

In a world context, Mark Spalding of The Nature Conservancy identified 232 'eco-regions' for coastal and shelf areas of the world of which two include the coastal shelf waters around Britain.

Major biogeographical regions around the British Isles.

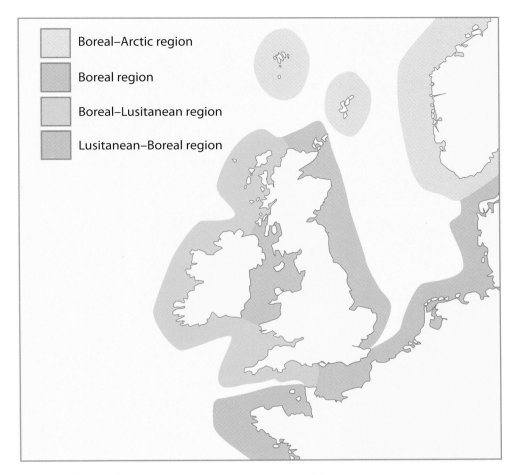

Boreal–Arctic region

Boreal region

Boreal–Lusitanean region

Lusitanean–Boreal region

Great Britain lies, as do many countries, across a biogeographical transition zone. The major biogeographical areas that are characterised by distinctly different assemblages of species were described by Edward Forbes and published in 1858. They are shown on the map reproduced in the chapter entitled 'Exploration and discovery: the first 150 years'. Those regions reflect a warmer or 'Lusitanian' region to the south of Britain, a 'Boreal' region which is centred on the British Isles and an 'Arctic' region much further to the north. For the British Isles, the shallow seas are described as 'Boreal–Lusitanean' along the western seaboard, 'Boreal' in the Irish Sea and off the eastern seaboard and 'Boreal–Arctic' around Shetland. Within the major divisions, there will be further sub-regional characteristics brought about by such features as water quality (in its broadest sense including turbidity and salinity), by temperature range through the year, by the predominant substratum type, by practical descriptors (such as 'The Irish Sea', The North Sea') and by characteristic coastal features such as the sea lochs of western Scotland. Sub-regions have been defined since the mid-1970s and redefined for different purposes – scientific and political. The sub-regions (often called 'Regional Seas') help us to assess biological character, including occurrences of species and biotopes. They also help us to compare the quality of those biotopes in terms of species richness or the presence of a rare or threatened species, and to compare habitats within biogeographically similar areas. There are between six and 15 sub-regions, depending on which authority you follow.

OCEAN CURRENTS

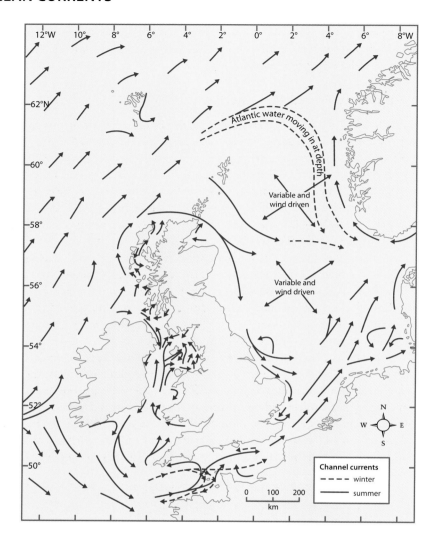

The direction of near-surface currents around the British Isles.

When the tidal component is removed from currents, there remains a residual flow that is important in the distribution of water masses and of larvae and spores and some adult low-mobility seabed species. The role of these currents is sometimes appreciated when, for instance, swarms of oceanic jellyfish arrive at the coast or when sea beans (seeds and fruits of tropical trees) are washed ashore.

WAVE ACTION

The training of marine ecologists usually involves visits to shores of different wave exposures. Jack Lewis's seminal book *The Ecology of Rocky Shores* (published in 1964) was structured, in part, around describing shores from wave-exposed to wave-sheltered conditions. In that book, he provided a Biological Exposure Scale Applicable to British Coasts perhaps spurred-on by the Biologically-defined Exposure Scale published by Bill Ballantine in *Field Studies* in 1961. Both Lewis and Ballantine were

Below left: we were so lucky! Rockall and its associated reefs (The Rock of Rockall and Hasselwood Rock are shown here in a picture with David Connor and the author in 1988) are subject to violent wave action but, on the one and a half days we had to survey them, there was no wind. Rockall is 370km west of North Uist and 306km west of St Kilda, isolated from the shallow seas around Britain by the Rockall Trough which reaches to depths of over 3,000m. Image: Dan Laffoley.

Below right: for most of the time (but not when we were there), the water is incredibly clear and foliose algae were found to extend to a depth of 45m, deeper than at St Kilda. The kelp *Alaria esculenta*, normally found only in the sublittoral fringe, is here photographed at 30m depth. The backscatter is from a plankton bloom. Image width *c*. 1m. Image: Christine Howson/JNCC. '*Alaria esculenta* forest with dense anemones and crustose sponges on extremely exposed infralittoral bedrock' (A3.112 / IR.HIR.KFaR. AlaAnCrSp).

rather sloppy in using the word 'exposed' and I often followed suit until brought to book by one Magnus Magnusson who enquired 'exposed to what?'. Duly chastened, I am now always careful to refer to exposure to wave action and exposure to tidal currents or just exposure to water movement. 'Exposure' to such things as emersion, to sand scour, to reduced salinity, etc. are other topics.

Ultimate exposure to wave action – Rockall

The reefs surrounding the rock of Rockall are no doubt off the scale of exposure to wave action used in biological surveys. This ultimately wave-exposed location was surveyed in June 1988 by a team from the Marine Nature Conservation Review (advised by divers from Inverness Sub Aqua Club who had already been there in 1987). Arranging such an expedition was no small feat and it was accomplished by one of the team members, Dan Laffoley. The expedition involved hitching a lift on the Fisheries Protection Vessel *Norna* and hiring, just in case, a recompression chamber duly welded to the deck. Remarkably, the sea was glassy calm, although a large swell prevented any survey work in depths shallower than 6m below sea level. Photography was also difficult in shallower depths but the broad picture was of shallow rocks dominated by the kelp *Alaria esculenta* and, on the rocks below, by encrusting calcareous algae, hydroids including the tubular hydroid *Tubularia indivisa*, and tubicolous amphipods (*Parajassa pelagica* to about 14m and *Jassa falcata* deeper). *Alaria*, which usually extends to depths of no more than 2m below low water level, extended to 33m at Rockall (it extends to 17m at St Kilda) and dominated many areas of rock. Under the kelp and on vertical surfaces the reef was colonised especially by Jewel Anemones *Corynactis viridis*, Elegant Anemones *Sagartia elegans*, the cold water anemone *Phellia gausapata*, encrusting sea squirts and encrusting sponges. Foliose red seaweeds were sparse and there were few species. Sea urchins, *Echinus esculentus*, grazed the rocks and were found as shallow as 8m below sea level – quite a feat of adhesion in such wave-exposed conditions. Rockall (but after a brief survey) was considered impoverished,

Under the kelp in moderate depths, rocks were characterised by a colourful array of Jewel Anemones *Corynactis viridis*, encrusting sea squirts, encrusting sponges, pink encrusting algae, a Bloody Henry starfish *Henricia* sp. and sea anemones including the cold water species *Phellia gausapata*. Image width *c.* 25cm. Image: Sue Scott/JNCC. Closest to: 'Foliose red seaweeds on exposed lower infralittoral rock' (A3.116 / IR.HIR. KFaR.FoR).

especially for seaweed species, compared to mainland sites and St Kilda: most likely a result of its isolation rather than extreme exposure to wave action. A survey report was written by Christine Howson and specimens deposited in the National Museum of Scotland in Edinburgh, but there is no paper published in the scientific journals.

At the deepest point, 46m, surveyed by diving on Helen's Reef, the steep rock slopes had turned to a boulder tumble colonised by Jewel Anemones, encrusting sponges and sea squirts, sparse Dead Men's Fingers *Alcyonium digitatum* and grazed by sea urchins *Echinus esculentus*. Here, at 43m depth. Image: Keith Hiscock/ JNCC. Closest to: '*Corynactis viridis* and a mixed turf of crisiids, *Bugula*, *Scrupocellaria*, and *Cellaria* on moderately tide-swept exposed circalittoral rock' (A4.132 / CR.HCR. XFa.CvirCri).

Ultimate shelter from wave action and tidal currents – Abereiddy Quarry

A group of lecturers and undergraduate diving biologists from Westfield College, University of London were spending a weekend diving in Pembrokeshire in November 1969. When a force 8 gale made diving at open coast sites impossible, a sheltered site was sought and Abereiddy Quarry was discovered. The quarry was flooded in the 1930s to provide a harbour and is 24m deep with a summer thermocline that isolates waters below about 12m depth. The quarry was one of my sites of ultimate shelter (the other was the Western Trough of Lough Hyne in Ireland) studied as a part of my PhD with an account published in the *Journal of the Marine Biological Association*. Together with Richard Hoare, I dived the quarry, taking samples and making records approximately every 10 weeks for over a year. I have revisited the quarry several times since the early 1970s and these photographs are from August 2016.

Shallow algae in such a wave and tide-sheltered situation are typically of various filamentous red species and stringy red weed *Polysiphonia elongata*, with Sugar Kelp *Saccharina latissima* and Bootlace Weed *Chorda filum*. Image width *c.* 1.5m. '*Saccharina latissima* on very sheltered infralittoral rock' (A3.313/IR.LIR.K.Slat).

In Abereiddy Quarry, shaded surfaces are characterised by the Double Spiral Worm *Bispira volutacornis* with solitary sea squirts *Ascidia mentula*, and the attached scyphistoma stage of the Moon Jellyfish *Aurelia aurita*, together with a variety of sponges typical of shaded surfaces. (In Scotland, very sheltered situations are likely to be characterised by another species of fan worm, The Peacock Worm *Sabella pavonina*.) Image width *c*. 20cm. Close to: 'Solitary ascidians, including *Ascidia mentula* and *Ciona intestinalis*, on wave-sheltered circalittoral rock' (A4.311 / CR.LCR.BrAs.AmenCio).

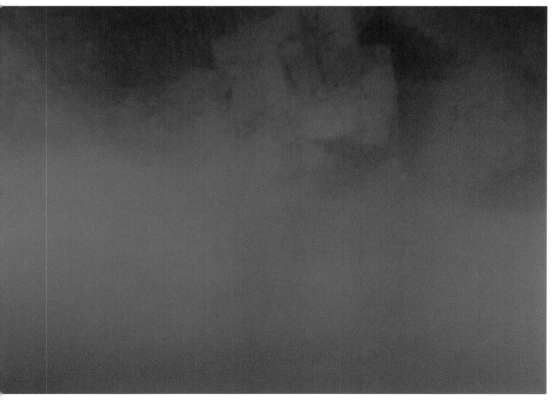

In such extreme shelter and depth as at Abereiddy, the still water means that a thermocline is formed from early summer until about November (when seawater temperatures on the open coast roughly equate to those below the thermocline). That thermocline isolates deeper water which becomes deoxygenated. The thermocline moves up and down a little and, here, the rocks above the cloudy water at the thermocline are bare of conspicuous life. Image width *c*. 50cm.

Extreme shelter from wave action and from tidal currents occurs in enclosed bays including some sea lochs and voes as well as man-made habitats such as Abereiddy Quarry and harbours that have been protected by walls and breakwaters. There are similar slate quarries to Abereiddy at Easdale, south of Oban on the Scottish west coast, that Trevor Norton surveyed for algae in the mid-1960s and that I surveyed in 1984. Here, the abundant sea squirts were Yellow-ringed Sea Squirt *Ciona intestinalis* and *Ascidiella aspersa*, *Aurelia* scyphistomae were again abundant but, typically for Scotland, there were also large numbers of the brachiopod *Novocrania anomala*. At Easdale in July 1984, there was no thermocline and no apparent deoxygenation but water temperature fell from 12.5°C in shallow depths to 7.7°C deeper than 24m. However, Trevor Norton describes (in the journal *Hydrobiologia*) that, in September 1966 and deeper than 24m, there was a black particulate 'soup' which smelt of hydrogen sulphide, suggesting anaerobic conditions.

The Double Spiral Worm *Bispira volutacornis* thrives in the shelter of Abereiddy Quarry. Image width *c.* 25cm.

Wave action is attenuated with depth. On open coasts facing into the ocean, gale force winds will cause apparently highly destructive multi-directional water movement to a depth of about 20m before becoming oscillatory. Below a depth of about 80m, strong wave action at the surface will hardly be felt. The swell waves that are created by distant storms often have a very long wavelength and penetrate much deeper than wind-driven waves.

The strength of water movement near the seabed under a fully-developed force 8 gale.

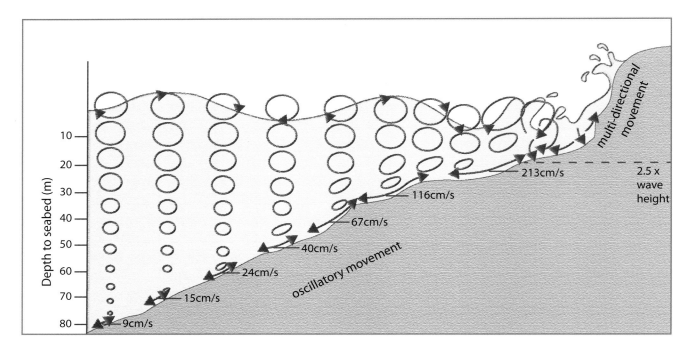

TIDAL CURRENTS

Britain has some of the largest tidal ranges in the world: up to 15m in the Severn Estuary. On the rising or falling tide, water moves between locations and, where the tidal excursion is large, the movement of water is likely to create strong tidal currents. If that water movement is squeezed between land masses or is blocked by an underwater obstruction, tidal flow accelerates, sometimes creating overfalls and whirlpools at the surface. In between, on the open coast, in bays or the inland areas of sea lochs, the strength of tidal flow may be negligible. Since the subject of my PhD thesis was the effect of water movement on the ecology of sublittoral rocky areas, I studied a wide range of locations subject to different degrees of tidal current strength. I was able to identify points along the continuum of strength of tidal current velocity where marked changes occurred in seabed biology. From this, a scale of exposure to tidal current strength was developed that could be applied in surveys and continues to be used today. The range of different exposures to tidal current velocity that might be experienced around a hypothetical island are illustrated below.

The richest reef communities may be found where tidal currents are moderate – the currents bring food for suspension feeders and keep rocks and organisms clear of silt whilst not creating mechanical stress.

Some effects of coastal topography on the velocity of surface tidal currents.

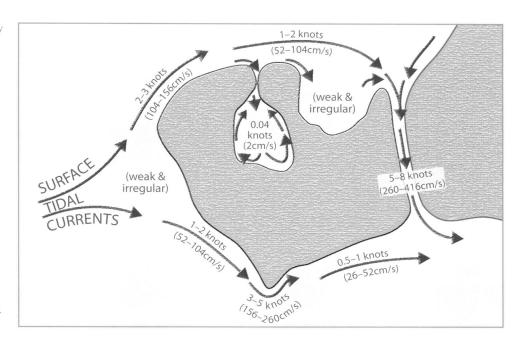

Below: deeper than vigorous and potentially damaging wave action and in moderate tidal currents (less than about 3 knots), rich communities of animals may develop. Here, a diver is surveying in Wembury Bay in south Devon. '*Eunicella verrucosa* and *Pentapora foliacea* on wave-exposed circalittoral rock' (A4.1311 / CR.HCR. XFa.ByErSp.Eun).

WAVE ACTION AND TIDAL CURRENTS TOGETHER

The strength of wave action and tidal currents, together with light intensity, are the major factors determining the rocky seabed communities that develop at particular depths. However, the strength of wave action is attenuated with depth, and the creation of a two-part exposure scale for wave action and tidal currents together, based on the communities present at a location, becomes a seeming impossibility. Explaining that dilemma to Bill Ballantine in the library at Orielton Field Centre in the late 1970s only led to what was basically an instruction to stop moaning and get on with working out how to do it. The resulting diagram was published in *Sublittoral Ecology: the Ecology of the Shallow Sublittoral Benthos* in 1983 and is slightly updated and summarised here.

Opposite: high energy environments, whether because of strong wave action or strong tidal currents, result in broadly similar assemblages of species visually dominated by tubular hydroids (here, *Tubularia indivisa*) and Elegant Anemones *Sagartia elegans* as well as encrusting sponges, Jewel Anemones *Corynactis viridis* and certain sea firs. The tubular hydroids live for only a short period in spring and the dead stalks may later become covered in the tubes of jassid amphipods. Part of the Eddystone reef in April. Image width *c.* 12cm. The image is from 18m depth and a variant of '*Corynactis viridis* and a mixed turf of crisiids, *Bugula*, *Scrupocellaria*, and *Cellaria* on moderately tide-swept exposed circalittoral rock' (A4.132 / CR.HCR.XFa.CvirCri).

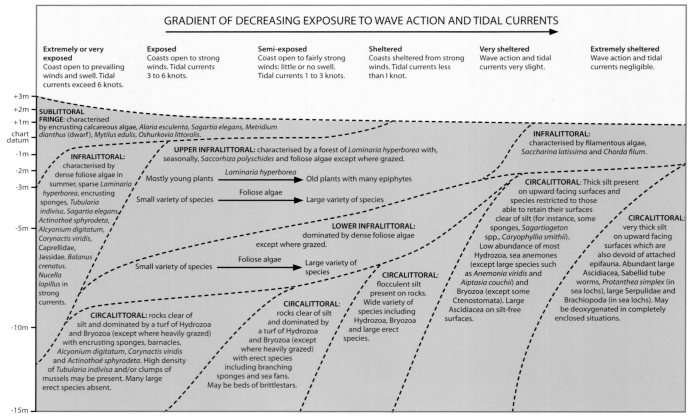

Changes in seabed communities with decreasing exposure. Similar communities develop in relation to strength of wave action and tidal currents except that shallow algal and animal assemblages are shaped mainly by the strength of wave action.

LIGHT AND DEPTH

The intensity of light penetrating the sea is rapidly reduced with increasing depth, and the spectral composition also changes. In some locations around Britain, turbidity is so high that there is insufficient light, even at the shallowest depths, for algae to grow. In others, the water is clear for most of the year and algae may dominate rocks to depths in excess of 25m in sheltered conditions and where there is no or little impact from urchin grazing. The deepest that multicellular algae have been recorded on the seabed around Britain is 90m. That record, in a paper by Julian Clokie and others, was from the Rockall Bank. In general, the critical depth below which kelp plants fail to grow is where about 1% of surface illumination is reached. For foliose algae, the critical depth is that at which about 0.1% of surface light penetrates. Shading by algae, especially kelps, is also important in determining what can grow under the canopy. It was Jack Kitching, working with primitive diving equipment on the west coast of Scotland from 1932 to 1936, who determined that the kelp canopy could cut out about 99% of light.

Opposite: kelp fronds can cut out up to 99% of the light that reaches the canopy.

Below: the change in dominant species on rock surfaces with increasing depth on the open coast is brought about mainly by attenuation of light. However, it is also influenced by grazing, which can reduce algal cover, and by strong wave action that can favour some sessile animal species over algae. The depths are typical for open coast areas in areas with low levels of turbidity.

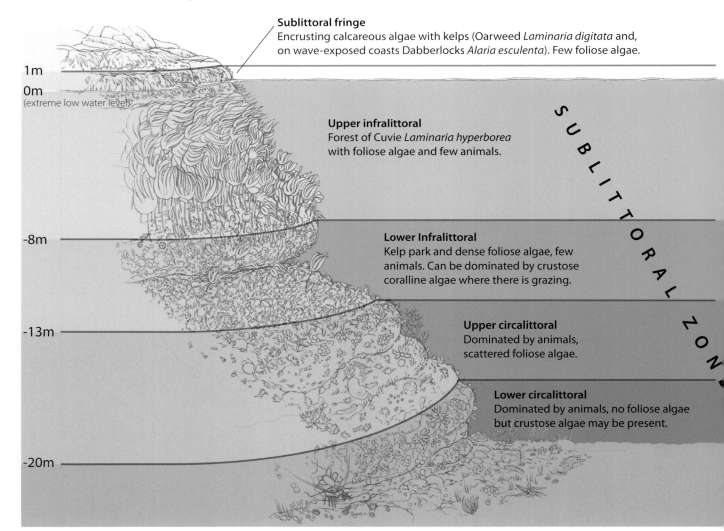

Sublittoral fringe
Encrusting calcareous algae with kelps (Oarweed *Laminaria digitata* and, on wave-exposed coasts Dabberlocks *Alaria esculenta*). Few foliose algae.

1m
0m
(extreme low water level)

Upper infralittoral
Forest of Cuvie *Laminaria hyperborea* with foliose algae and few animals.

-8m

Lower Infralittoral
Kelp park and dense foliose algae, few animals. Can be dominated by crustose coralline algae where there is grazing.

-13m

Upper circalittoral
Dominated by animals, scattered foliose algae.

Lower circalittoral
Dominated by animals, no foliose algae but crustose algae may be present.

-20m

SUBLITTORAL ZONE

Measuring the deepest extent of foliose algae at Lundy. Here, in September 2010 when, corrected for depth below chart datum and calibration of the diving computer, it was 24.5m. In 1985 and 1986, the deepest foliose algae had been recorded at about 22m below chart datum. The alga is Polkadot Weed *Haraldiophyllum bonnemaisonii*.

There are 'confusing factors' in the simple story of light attenuation with depth causing a change from algal domination to domination by animals. In areas of very strong water movement (wave action or tidal currents), the food supply for suspension-feeding animals is so abundant that they may out-compete algae for space. Urchin grazing has a very significant effect on reducing the downward extent of macroalga, especially the kelps.

Light is essential for photosynthesis and some animals such as the Snakelocks Anemone *Anemonia viridis*, that have algae associated with their tissue, thrive only in shallow water. Image width *c.* 12cm.

SALT AND FRESH

The character of seabed communities on open coast and offshore areas is unlikely to be greatly affected by variable or reduced salinity except where that coast is near to the outflow of a major river such as the Severn or the Humber. In general, communities that are characteristic of fully marine conditions occur where salinity is over 30 (the number refers to practical salinity units). Nevertheless, there may be species that only thrive in fully saline and unvarying conditions of salinity, but whether it is salinity or some other factor such as turbidity that affects them is likely to be unclear. Along marine inlets with freshwater input, salinity will decline or become variable with increasing distance from the open coast and there will be consequential changes in the seabed communities present. Those changes are explained further in the 'Estuaries' section.

WEATHER EVENTS

Weather events might include extremely cold winters or warm summers, heavy persistent rainfall and severe storms. The very cold winter of 1962/63 (I was at school, where we enjoyed winter sports that seemed greatly to involve broken limbs and a lot of bruising) affected especially intertidal species, but also some subtidal organisms. Denis Crisp catalogued, in the *Journal of Animal Ecology*, the many observations that were made around Britain, including events for subtidal species. Fish were found dead or in a torpid state, including strandings of wrasse and Conger Eels. Spiny Spider Crabs *Maja brachydactyla* and European Spiny Lobsters *Palinurus elephas* died and there were mass strandings of burrowing sea urchins *Echinocardium cordatum* and of the starfish *Astropecten irregularis*. There was a high mortality of oysters and, in the same areas on the east coast, of Peacock Worms *Sabella pavonina*. In the concluding

'Wash-outs' of mobile species, sediment infauna or of species attached to pebbles and shells in the sediment are a frequent occurrence after storms. Here, on the shore at Hunstanton in October. Image width *c.* 20cm.

Following vicious storms in the winter of 2013/14 we returned to the reefs offshore of Plymouth near to where sea state had reached 'Phenomenal'. Although somewhat silty and with a fog of very fine sediment particles in the water, the marine life on reefs looked much as always. This image was taken at Hand Deeps on 15 March 2014. Image width *c.* 50cm.

comments, Crisp states: 'As expected the heaviest mortalities were found mainly amongst the southern or Lusitanian elements of the British fauna'. On the other side of the coin, warm weather events seem to attract less attention or, perhaps, they are a rarity in Britain. Extreme elevated temperatures in other parts of Europe (for instance in the Ligurian Sea, north-west of Italy) have caused mortalities in a range of seabed invertebrate species.

From mid-December 2013 to the end of February 2014, the coasts of south-west and southern England and Wales were battered by a succession of severe gales, introducing a new category of sea state to my vocabulary: 'Phenomenal'. It might be assumed that those severe storms would have caused significant re-arrangement of seabed habitats and associated assemblages of species. Some sediments were moved around and scallop fishermen off south Devon complained that what had been level seabed now had boulders exposed. Burrowing sea cucumbers and some sea anemones (*Mesacmaea mitchellii*) were being caught in trawls, having been displaced from sediments and, it seemed, were unable to re-burrow. There was a spectacular wash-out of Common Otter Shells *Lutraria lutraria* in Whitsand Bay west of Plymouth, together with localised strandings of razor fish, *Ensis* sp.(p). Seagrass beds were damaged in places. Indeed, some species that we didn't know were there, such as mantis shrimps *Rissoides desmaresti* at Dungeness in Kent and at Felpham in West Sussex, were washed out of sediments. There were some strandings of a few reef species such as Pink Sea Fans *Eunicella verrucosa*. When we returned to diving out of Plymouth a couple of weeks after the storms abated, I expected to see detached species strewn about on the seabed or washed into gullies, but there was virtually nothing like that in the region of reef habitats: the attached marine life was much as always.

MODIFYING FACTORS

Knowing the strength of wave action and tidal streams, you should be able to predict sediment type and therefore biotopes present? No – and such a simple view has made many a model produce incorrect or misleading results. For instance, in parts of the Severn Estuary, the amount of riverborne suspended sediment in the water is so high that, even in very strong tidal currents, the seabed is of mud. There are many places around the coast of Britain where there are legacy (coarse) sediments reflecting past geological events such as the moraines from glaciers or the coarse gravels of long-since inundated river beds and deltas (see the 'Sediments' section at the start of the following chapter on 'Habitats').

Knowing the clarity of the water on rocky coasts, you should be able to predict the depth to which kelp and foliose algae will penetrate? No – where wave action or tidal currents are strong, suspension feeding animals will be so successful that they dominate shallow rocks. No – where sea urchin grazing occurs, there will be urchin barrens below the kelp and the kelp will not penetrate to depths that light would allow.

Knowing the direction and strength of residual currents, you should be able to predict 'connectivity distances' for seabed species – where larvae and spores from one location are likely to settle in another? No, not necessarily. The naïve scientist will use models that predict direction of travel of inert particles – formulae borrowed from coastal geomorphologists. Many seabed species have very poor powers of dispersal – they are fixed or of low mobility and their larvae and spores travel only a very short distance from their parents. The larvae of many seabed species that do travel significant distances are more clever than inert particles and may seek out particular conditions including the presence of others of their species, a substratum that will support them and clues such as the sound of crashing waves.

PHYSICAL AND BIOLOGICAL FACTORS: BRINGING IT ALL TOGETHER

The diagram on page 58 brings together, as far as possible, the physical and chemical (salinity) conditions, together with substratum type and biology that determine what species occur where. It was drawn to support a research report on 'The Structure and Functioning of Marine Ecosystems' prepared for English Nature in 2006. To understand what factors are important to the occurrence and the survival of species and communities at a location, we need to take account of at least the following properties and processes (which are in alphabetical order; importance will vary from location to location):

- Absorption of gases (especially oxygen)
- Contaminants – type and levels
- Dispersal processes for larvae and propagule and for adults (especially the role of residual currents but also of larval or adult behaviour and their responses to environmental 'clues' of where to go)
- Grazing
- Microbial processes
- Migration
- Nutrients, exchange and supply

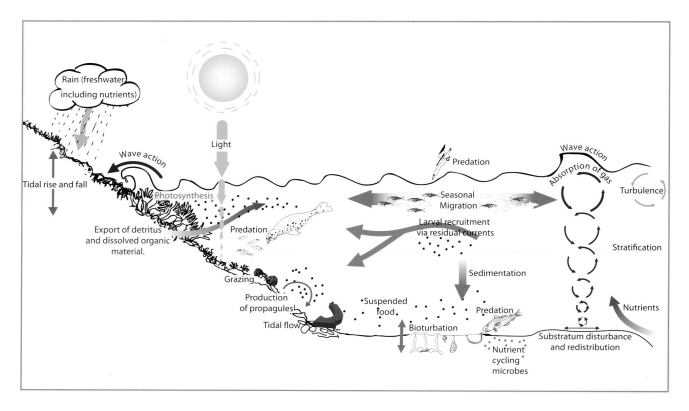

Physical, chemical and biological processes that influence shallow seabed communities on the open coast. Image: Keith Hiscock / Jack Sewell.

- Nutrient recycling, especially the role of microbes and of bioturbation
- Photosynthetic processes including amount and type of light available at the seabed
- Predation
- Residual current strength and direction
- Salinity and salinity discontinuities with depth
- Sediment disturbance and redistribution (especially by storms)
- Sediment supply and sedimentation processes
- Substratum type including structural features (caves, etc.)
- Supply of food (suspended or seabed)
- Temperature range
- Temperature discontinuities (laterally and with depth: thermal stratification)
- Tidal current strength
- Turbidity
- Water quality characteristics brought about by movement of different water masses
- Wave action (strength and type).

The wide range of conditions combine to determine what seabed communities occur where, and come together in the major physiographic features around the coasts of Britain. The next diagram is based on one that was drawn to illustrate major physiographic and other features that hold distinctive seabed habitats. The original diagram provided one of the ways in which sites were categorised in the Marine Nature Conservation Review of Great Britain.

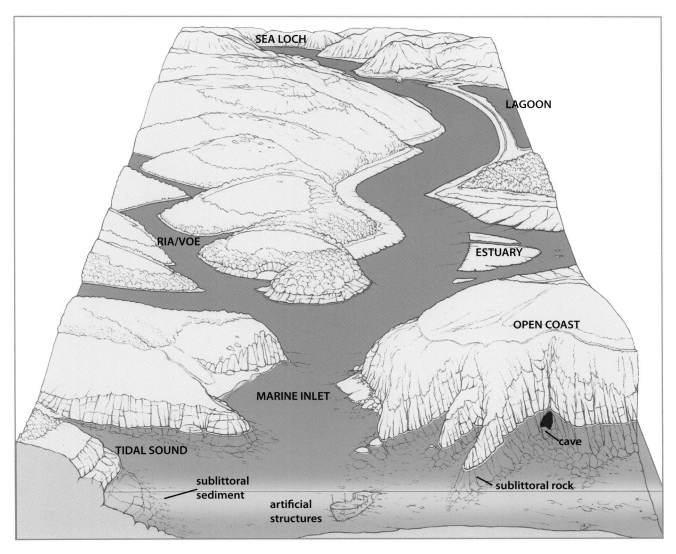

Coastal physiographic and habitat features likely to hold distinctive communities of species. Based on a diagram in the *MNCR Rationale and Methods*.

HABITATS

SEDIMENTS

This section concentrates on what can be seen at the surface of the sediment and how that relates to what is below. It benefits from line drawings by Jack Sewell giving an impression of what lives where in the sediment. It draws examples mainly from areas of inshore open coast. Sediment communities and species that seem particular to, or especially found, in physiographic features such as rias (including voes), sea lochs and estuaries are described in separate sections on those physiographic features.

Above and below the sediment surface

Most of the rich diversity of animal species in subtidal sediments is hidden from view beneath the surface. Occasional signs of life such as burrows, feeding siphons, tubes, tentacles of various sorts and the arms of brittlestars give only clues as to what lives below. Whilst level seabeds can be surveyed by cameras mounted on towed sledges or, more cleverly, arrays that stay just above the seabed, only species living on the surface will be revealed. More often, dredges and grabs are used to collect samples that are then sieved over a mesh (usually of 1mm or 0.5mm) so that the species living in the sediment can be identified. Even then, only the upper 10 or maximum 20cm of sediment is sampled, and some species live deeper than that.

The contents of a 0.1m² grab sample with sediment sieved out over a 1mm mesh from a rich patch of sandy mud in Liverpool Bay. There are large numbers of tubes of the Trumpet Worm *Lagis koreni* and of the White Furrow Shell *Abra alba*, together with burrowing sea urchins *Echinocardium cordatum*, razor shells *Phaxas pellucidus* and brittlestars *Ophiura albida* and possibly *Ophiura ophiura*. Image width: *c.* 35cm. Image: Ivor Rees. '*Lagis koreni* and *Phaxas pellucidus* in circalittoral sandy mud' (A5.355 / SS.SMu.CSaMu.LkorPpel).

The great majority of the continental shelf off Britain is composed of sediments. That shelf is quite wide, extending to over 160 kilometres in several places, before reaching the continental slope at depths of around 200 metres. The character of seabed sediments is determined by the nature of the source material, by the action of waves and by the strength of tidal currents, but also by history. A major influence on the medium scale topography and the origin of much sediment relates to sea level changes during the Pleistocene glaciations. At the peak of the glaciations, sea levels were over 100m below present and so much of the seabed was subjected to surf zone erosion and re-working of deposits from the glaciations. This includes the large amounts of glacial till in areas roughly north of the limits of the ice sheets (south Wales in the west, Norfolk in the east) and areas of seasonal outwash to the south. The extensive area of glacial moraine (boulders, cobbles and pebbles) that constitute the Sarns off Harlech in west Wales are colonised by reef species – most conspicuously, algae. The course of rivers can still be traced across the now drowned landscape and those rivers carried gravel and other coarse sediments that are extracted today, especially as a source of building materials. The greatest effect of the ending of the last ice age was in the Dover Strait, where a huge ice-dammed lake in the Southern North Sea broke through a chalk ridge. This caused a major flood spate whose effects are still visible in the topography of the Eastern English Channel today.

The character of seabed sediments has been mapped since the early days of hydrographic surveying, when a blob of tallow on the sounding lead would collect a surface sample. Later, grabs would be used to collect samples for geological surveys, but those surveys had two major drawbacks that produced misleading information. If the grab came up with nothing, the bottom might be mapped as 'rock', or the grab deployed until it eventually brought up sediment – perhaps from an isolated patch – and that sediment type would be mapped as representative. Furthermore, boulders, cobbles and pebbles (that may be colonised by hard substratum species) were mapped as 'gravel' sediments. We have been able to do better at identifying coarse sediments since the use of cameras and diving, but the sediment classification used by geologists continues to stop at gravel.

Level sediment habitats are home to worms, small crustaceans, bivalve molluscs and other species that are an important food source for fish that live on the bottom or close to it. In turn, many species of fin fish including Plaice, Cod, Haddock, Whiting, Sole, Red Mullet are caught by trawling. Sandeels that live in the sediment are an important part of the diet of Turbot, seals and seabirds. Shellfish that live in subtidal sediments and that are commercially exploited include cockles, razor clams, native oysters, scallops and, in deep mud areas, Scampi or Norway Lobster.

Plaice *Pleuronectes platessa* have evolved to lie flat on the seabed where they feed on worms, small crustaceans and the siphons of bivalve molluscs. Such bottom-feeding fish are an important part of the sediment seabed community. Image width *c.* 25cm.

Many surveys of sediment communities have been carried out, and some of the widescale ones are mentioned in the earlier chapter on 'Gathering knowledge and creating information'. As an example of the range of different sediment types and associated fauna that may be seen in a fairly small area, four images from Caernarfon Bay in north Wales have been selected for illustration.

Level seabeds may have a variety of sediment types. Here, in Caernarfon Bay, they range from muddy and sandy sediments colonised by out-of-sight burrowing species to ones with coarser material colonised by epifauna. Photographed as part of the HABMAP project (National Museum of Wales BIOMÔR 5). Image widths are about 40cm. Images: Aquafact International Services Ltd, Galway.

Rippled muddy sand with a cast of the Lugworm *Arenicola marina*. [BIOMÔR 5 Stn 72]. Most likely '*Arenicola marina* in infralittoral fine sand or muddy sand' (A5.243 / SS.SSa.IMuSa.AreISa).

Muddy sand with shells colonised on the surface by brittlestars *Ophiura ophiura*. [BIOMÔR 5 Stn 74]. 'Circalittoral muddy sand' (A5.26 / SS.SSa.CMuSa) and sampled near to sites characterised as '*Lagis koreni* and *Phaxas pellucidus* in circalittoral sandy mud' (A5.355 / SS.SMu.CSaMu.LkorPpel).

Hard substratum (probably cobbles) overlain by coarse sand and colonised most conspicuously by Hornwrack *Flustra foliacea*, Ross 'coral' *Pentapora foliacea*, a branching bryozoan *Cellaria* sp. and an antenna hydroid *Nemertesia antennina*. [BIOMÔR 5 Stn 37]. Most likely '*Flustra foliacea* and *Hydrallmania falcata* on tide-swept circalittoral mixed sediment' (A5.444 / SS.SMx. CMx.FluHyd).

Cobbles and pebbles colonised by Dead Men's Fingers *Alcyonium digitatum* with Common Brittlestars *Ophiothrix fragilis* and pink encrusting algae. [BIOMÔR 5 Stn 25]. Biotope identity uncertain but near to areas occupied by '*Modiolus modiolus* beds with hydroids and red seaweeds on tide-swept circalittoral mixed substrata' (A5.621 / SS.SBR.SMus.ModT).

It is the grade (from coarse gravel to mud and clay) of sediment and the mixture of different sediments (for instance as muddy gravel, muddy sand, sandy mud, etc.) that determines what can live and thrive on and in the sediment. The amount of mud in a particular sediment is important in relation to how cohesive that sediment is likely to be: sediments that 'hold together' (consolidated sediment composed, by volume, of 20% or more of mud) will support permanent burrows. The degree of 'bioturbation' is also important: species that dig, burrow and plough through sediment

Rich communities in undisturbed sediments

The seabed around the Skomer Marine Nature Reserve (now Marine Conservation Zone) has been protected from mobile fishing gear for more than 25 years. Over the course of monitoring the various sediment types, over 1,000 species have been recorded on and in them at Martin's Haven and along the adjacent north-facing coast of the Marloes Peninsula. Many of the sediments correspond to 'Cerianthus lloydii with Nemertesia spp. and other hydroids in circalittoral muddy mixed sediment' (A5.4411 / SS.SMx.CMx.ClloMx.Nem).

Left: muddy sand sediments in Martin's Haven are colonised by many burrowing species, most conspicuously anemones. Here, two Burrowing Anemones *Cerianthus lloydii* and a Policeman Anemone *Mesacmaea mitchellii*. Image width *c.* 25cm.

Bottom left: a 'natural' abundance and size range of King Scallops *Pecten maximus* can now be seen, providing an indication of what an unexploited seabed looks like. Here, a scallop and Auger Shells *Turritella communis*. Image width *c.* 40cm.

Bottom right: in shallow well-lit waters with coarse material (shells and pebbles) below the surface, algae grow. In the centre of the image, the Circular Crab *Atelecyclus rotundatus*. Image width *c.* 30cm.

help to oxygenate it. Supply of organic material influences the availability of food and may come especially from faecal pellets of zooplankton or seabed species such as suspension-feeding benthos. Where a range of sediment types occurs in an area, the number of different species is likely to be high. For instance, samples of a variety of different sediments around Lundy, taken by Phil Smith and Rob Nunny in 2007, revealed 478 species from 49 survey stations. Particular biotopes are reasonably easily identified from survey data that includes samples of infauna but with much more difficulty or not at all from species that can be seen on the surface. Those surface-visible species are, for example, species such as burrowing sea anemones, crabs, whelks, scallops and brittlestars.

Coarse mobile sediments

Coarse mobile sediments, especially those subject to swell waves or to strong tidal currents, are often impoverished. They contain a few juvenile worm species and some bivalve molluscs that can re-burrow if disturbed, and are patrolled by crabs looking for carrion. The wave action creates sediment waves.

Sand waves off the wave-exposed coast of north Cornwall: here, in a depth of about 20m at The Quies near Newquay. Spiny Spider Crabs *Maja brachydactyla* and hermit crabs *Pagurus bernhardus* may patrol these sediments looking for carrion. Burrowing bivalves (visible by their siphons), most likely razor shells, *Ensis* sp., were seen in the sediment at this location. Image width *c.* 80cm. 'Circalittoral coarse sediment' (A5.14 / SS.SCS.CCS).

More stable waved shell and stone gravel may be populated by burrowing Gravel Sea Cucumbers *Neopentadactyla mixta,* as photographed here off Blackstone Point, south Devon. The sea cucumbers have a long muscular body buried in the sediment. The arms catch food passing by in the water and each is sequentially extended into or wiped across the mouth – divers should take a moment to watch. Image width *c.* 12cm. 'Neopentadactyla mixta in circalittoral shell gravel or coarse sand' (A5.144 / SS.SCS.CCS.Nmix).

Coarse shell gravel at the Mewstone near Plymouth. Wave action and strong tidal currents keep the sediment clean of silt, at least near its surface. This is an attractive habitat for sandeels, *Ammodytes* sp. that can dive into the sediment if spooked. Image width *c.* 1m.

Coarse stable sediments

Where sediment supply is of coarse material including shell fragments and is, at least at its surface, kept clean of silt, particular species are likely to be seen. These include sandeels, burrowing sea cucumbers and certain species of sea anemone. Within the sediment, species that need well-oxygenated conditions, such as burrowing sea urchins, may thrive.

Above: burrowing sea urchins are usually hidden beneath the surface of sediment. Here, in the same location as the previous image, a juvenile Purple Heart Urchin *Spatangus purpureus* ploughs through the shell gravel. Image width *c.* 10cm.

An intriguing species that lives in clean, coarse shell gravel is the lancelet *Branchiostoma lanceolatum*, here photographed on the gravel from which it was sampled. The species is normally buried with the anterior end (right hand side of the image) at the sediment surface. Lancelets are primitive chordates that help in tracing how vertebrates have evolved and adapted. Image width *c.* 8cm.

The burrowing Clock Face Anemone *Peachia boeckii* is found in a variety of sediments but especially stable coarse sand and gravel or even amongst pebbles. Strome Narrows, Loch Carron. Image width *c.* 8cm.

The Turban Top Shell *Gibbula magus* is found off western and northern coasts on gravel substrata. This one was photographed at Silver Steps near Falmouth. Image width *c.* 3cm.

Muddy or sandy mixed sediments

Coarse stable sediments are often the richest sediments for hidden infauna and for epifauna. They can be silty and may even appear to be mud because of a veneer of silt that settles in calm conditions. The sediment illustrated below, from off the east side of Lundy, yielded 217 taxa from seven grab sample stations in a sampling programme undertaken by Phil Smith and Rob Nunny for English Nature in 2007. Where coasts are sheltered from the waves (east-facing coasts, bays, some tidal sounds) and there is some tidal flow, sediments often provide stable conditions, with a supply of suspended food and therefore rich communities.

Coarse sandy gravel sediments about 1km off the east coast of Lundy and at a depth of about 20m. Only a few burrowing species show themselves at the surface – here a Policeman Anemone *Mesacmaea mitchellii* – while mobile species such as this Spiny Starfish *Marthasterias glacialis* seek out bivalve molluscs. Image width *c.* 20cm. Most likely '*Mediomastus fragilis, Lumbrineris* spp. and venerid bivalves in circalittoral coarse sand or gravel' (A5.142 / SS.SCS.CCS. MedLumVen).

Looks like mud but the mud is a veneer over gravelly sediment. The Daisy Anemones *Cereus pedunculatus* are attached to stones or shells below the surface off the east coast of Lundy. Image width *c.* 30cm. Closest to: '*Abra alba* and *Nucula nitidosa* in circalittoral muddy sand or slightly mixed sediment' (A5.261 / SS.SSA.CMuSa. AalbNuc).

Inhabitants of muddy gravel

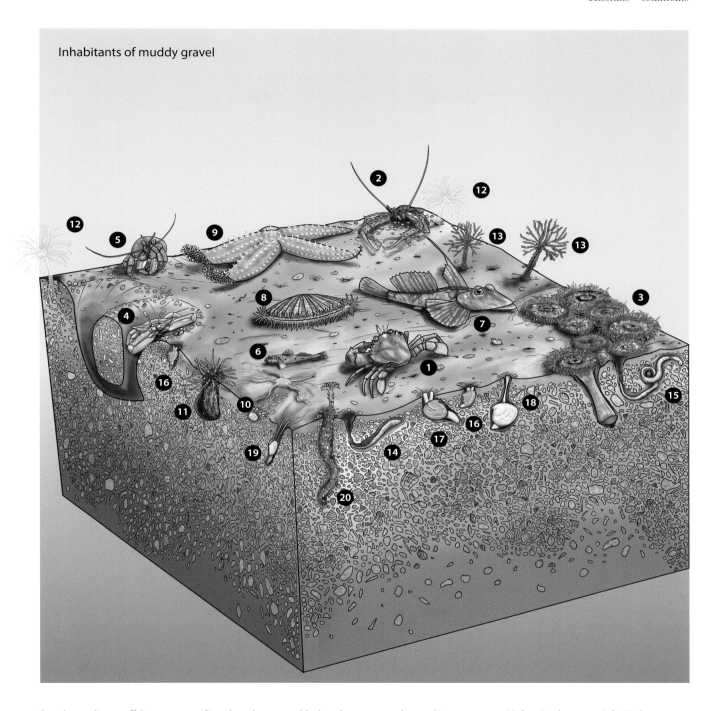

Based on sediment off the east coast of Lundy and more muddy than the coarse sandy gravel image on page 68. Species shown are **1** the Harbour Crab *Liocarcinus depurator*, **2** the Rugose Squat Lobster *Munida rugosa*, **3** a group of Daisy Anemones *Cereus pedunculatus*, **4** the Angular Crab *Goneplax rhomboides*, **5** a hermit crab *Pagurus bernhardus*, **6** sand gobies *Pomatoschistus* sp., **7** the Common Dragonet *Callionymus lyra*, **8** the King Scallop *Pecten maximus*, **9** the Spiny Starfish *Marthasterias glacialis*, **10** the brittlestar *Ophiura ophiura*, **11** the Policeman Anemone *Mesacmaea mitchellii*, **12** a solitary hydroid *Corymorpha nutans*, and **13** the Sand Mason Worm *Lanice conchilega*. Also burrowing in the sediment are **14** polychaete worms *Hilbigneris gracilis*, and **15** *Glycera lapidum*, **16** the bivalve mollusc *Timoclea ovata*, **17** *Venus thomassini*, **18** *Dosinia lupinus*, **19** *Gari tellinella* and **20** a tube worm *Owenia fusiformis*. Several species of polychaete worms and amphipod crustaceans, also characteristic of this habitat, are not shown. Close to '*Mediomastus fragilis, Lumbrineris* spp. and venerid bivalves in circalittoral coarse sand or gravel' (A5.142 / SS.SCS.CCS.MedLumVen). Drawing: Jack Sewell.

Some species that live in muddy or sandy mixed sediments and can be seen at the surface:

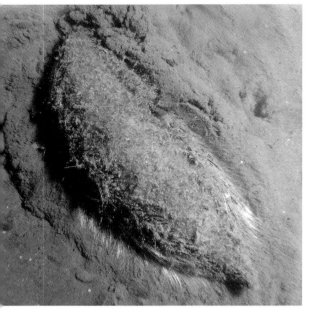

Left: a Sea Mouse *Aphrodita aculeata*, Plymouth Sound. The iridescent lateral chaetae (hairs) are characteristic. Image width *c.* 7cm.

Right: the hydroid *Corymorpha nutans* appears in the spring and may be abundant or hardly present at the same location in different years. Helford River. Image width *c.* 5cm.

Left: the solitary and unattached sea squirt *Molgula occulta* is a rarely encountered species in gravelly sediments. Knoll Pins at Lundy. Image width *c.* 4cm.

Right: fine tentacles of a terebellid worm that creates a tube of coarse sediment. These worms are widespread in sheltered, slightly muddy mixed sediments but their identity has yet to be determined. Firestone Bay, Plymouth Sound. Image width *c.* 8cm.

Left: the Rugose Squat Lobster *Munida rugosa*, here off the east coast of Lundy. This lobster lives in holes in a wide range of sediments as well as in rock fissures. Image width *c.* 8cm.

Right: the Pink Shrimp *Pandalus montagui*. Firestone Bay, Plymouth Sound. Image width *c.* 5cm.

Sand and muddy sand

Sandy sediments are widespread and occur in shallow depths in enclosed bays and inlets as well as in deeper water offshore. They are colonised by many species that occur in different biotopes. The illustration on page 73 shows components from several examples of what were previously described as the 'offshore muddy sand association'.

Some brittlestars, such as *Amphiura filiformis* and *Amphiura chiajei*, are infaunal (they live in the sediment) and are mostly only seen in grab or dredge samples. From time to time, they may be observed extending their arms into the overlying water to feed. Although illustrated here in Mounts Bay and the Helford River, burrowing brittlestars are a major component of many offshore sediments.

Shallow sand may be highly mobile and support only short-lived or juvenile burrowing species, those capable of re-burrowing, and mobile species such as fish and crustacea. The most conspicuous surface feature on shallow sandy sediments may be casts of the Lugworm *Arenicola marina*.

Shallow muddy sand, a widespread habitat, off Beesands in south Devon. Bivalve siphons, probably of *Ensis* sp. can be seen. Image width *c.* 25cm. Most likely '*Echinocardium cordatum* and *Ensis* spp. in lower shore and shallow sublittoral slightly muddy fine sand' (A5.241 / SS.SSa.IMuSa.EcorEns).

Below: the arms of burrowing brittlestars in muddy sand in Mounts Bay, Cornwall. Image width *c.* 35cm. Most likely '*Amphiura filiformis*, *Kurtiella bidentata* and *Abra nitida* in circalittoral sandy mud' (A5.351 / SS.SMu.CSaMu. AfilKurAnit).

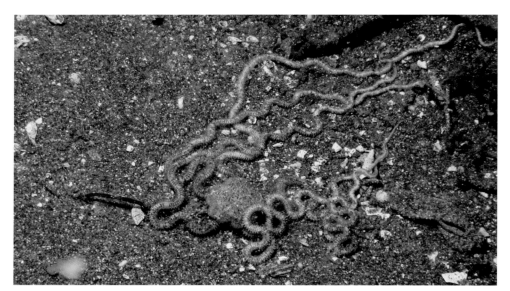

One of the burrowing brittlestars removed from the sediment. This one was identified as *Amphiura chiajei*. From Durgan in the Helford River. Image width *c.* 7cm.

Casts of the Lugworm *Arenicola marina* on shallow sandy sediments. Image width: *c.* 40cm. The biotope is most likely '*Arenicola marina* in infralittoral fine sand or muddy sand' (A5.243 / SS.SSa. IMuSa.AreISa).

Shallow sandy sediments that are colonised by seagrass *Zostera marina* are likely to hold a richer associated fauna than sediments that are not consolidated by the rhizomes of these flowering plants. Photographed off Ramscliff Point in Plymouth Sound. Image width *c.* 40cm. '*Zostera (Zostera) marina* beds on lower shore or infralittoral clean or muddy sand' (A5.5331 / SS.SMp.SSgr.Zmar).

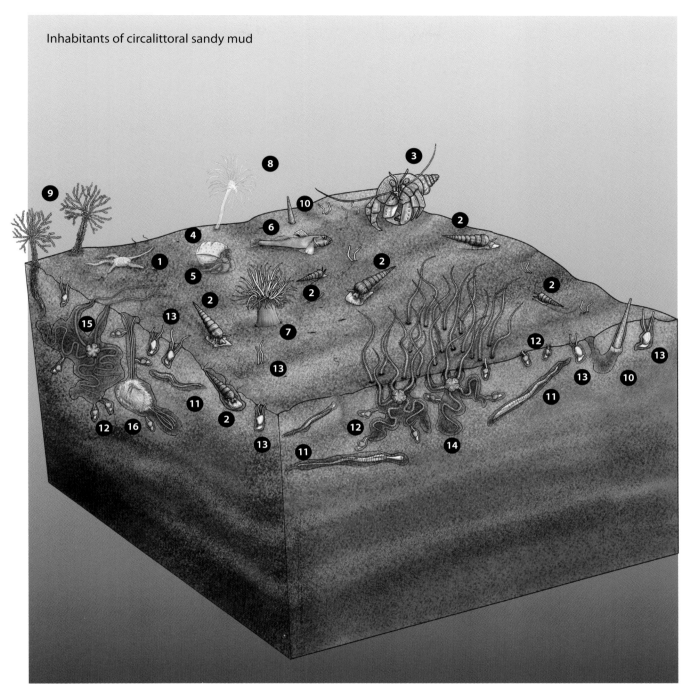

Inhabitants of circalittoral sandy mud

These species are present in widespread communities found in locations ranging from shallow waters in wave-sheltered inlets through to deep areas of offshore seabeds. Here, the species include, on the surface, **1** a brittlestar *Ophiura ophiura*, **2** Turret Shells *Turritella communis*, **3** a hermit crab *Pagurus bernhardus*, **4** the Cloak Anemone *Adamsia palliata* on **5** a hermit crab *Pagurus prideaux* and **6** sand gobies *Pomatoschistus* sp. Living in the sediment but with conspicuous structures above are **7** the anemone *Sagartiogeton undatus*, **8** the solitary hydroid *Corymorpha nutans*, **9** Sand Mason Worms *Lanice conchilega* and **10** the Trumpet Worm *Lagis koreni*. Burrowing in the sediment are **11** polychaete worms *Nephtys incisa*, **12** bivalve molluscs *Kurtiella bidentata* and **13** *Abra nitida*, **14** brittlestars *Amphiura filiformis* and **15** *Amphiura chiajei*. **16** The Sea Potato *Echinocardium cordatum* may be present in shallow depths especially. Close to 'Amphiura filiformis, Kurtiella bidentata and Abra nitida in circalittoral sandy mud' (A5.351 / SSS.SMu.CSaMu.AfilKurAnit) but including elements of other similar biotopes. Drawing: Jack Sewell.

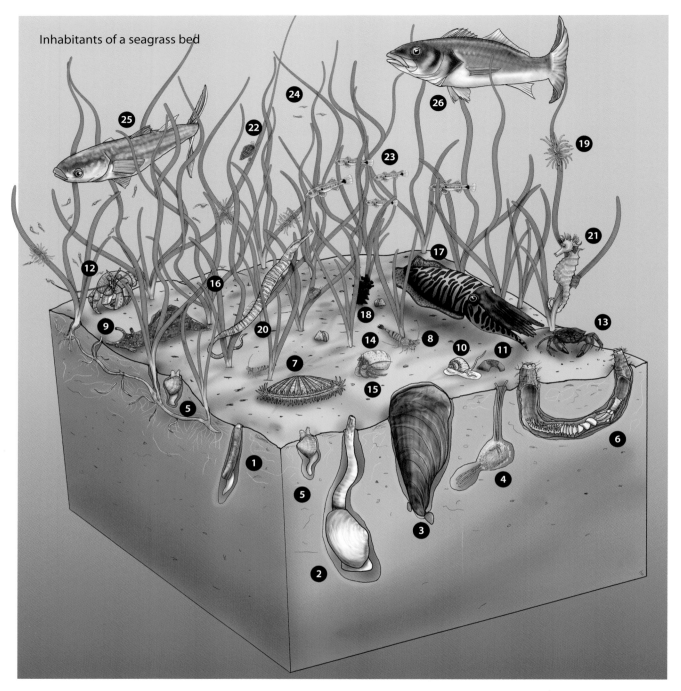

Inhabitants of a seagrass bed

Burrowing animals include **1** razor shells *Ensis ensis*, **2** the Common Otter Shell *Lutraria lutraria*, **3** the Fan Mussel *Atrina fragilis*, **4** the Sea Potato *Echinocardium cordatum*, **5** a bivalve mollusc *Cerastoderma edule*, and burrowing worms, most of which are inconspicuous but, here, represented by **6** the large tube-building parchment worm *Chaetopterus variopedatus*. On the seabed are **7** a King Scallop *Pecten maximus*, **8** a Brown Shrimp *Crangon crangon*, **9** a Sea Hare *Aplysia punctata*, **10** the Necklace Shell *Euspira catena* with **11** its eggs, **12** hermit crab *Pagurus bernhardus*, **13** the Shore Crab *Carcinus maenas*, **14** the Cloak Anemone *Adamsia palliata* on **15** a hermit crab *Pagurus prideaux*. Amongst the **16** seagrass *Zostera marina* fronds, **17** the Common Cuttlefish *Sepia officinalis* and **18** its eggs, **19** Snakelocks Anemones *Anemonia viridis*, **20** the Snake Pipefish *Entelurus aequoreus*, **21** the Long-snouted Seahorse *Hippocampus guttulatus*, **22** the Netted Dog Whelk *Tritia reticulata*, **23** Two-spot Gobies *Gobiusculus flavescens* and **24** mysid shrimps. Over the bed, **25** a Thick-lipped Grey Mullet *Chelon labrosus* and **26** a Sea Bass *Dicentrarchus labrax*. Drawing: Jack Sewell.

Mud and muddy sand

The muddy sediments of some very enclosed bays and deep offshore areas, below the significant effect of wave action, often contain rich and sometimes particular communities. Some characteristic ones have been mentioned in descriptions of physiographic features. It is the stability of these sediments that creates an environment encouraging the development of rich communities: the longer a habitat remains stable, the more different species are likely to settle there. The 'downside' of living in enclosed areas is that sediments are often polluted by industrial contaminants, sewage, anti-fouling paints, oil, road runoff (especially polychlorinated biphenols – from tyres) and thermal effluents. In addition, they are disturbed by navigational dredging, by the chains of boat moorings or navigation markers and by the presence of marinas, including importation of non-native species through shipping.

Mud habitats commonly occur in the shelter of breakwaters where tidal currents are minimal or in wave-sheltered bays such as Torbay. Here, the Angular Crab *Goneplax rhomboides*, one of the most frequent inhabitants of mud, but one of the most rarely seen on the surface, is caught unawares outside its burrow north of the breakwater in Plymouth Sound. Image width *c.* 30cm. Closest to: 'Seapens and burrowing megafauna in circalittoral fine mud' (A5.361 / SS.SMu.CFiMu.SpnMeg). (A very few Slender Sea Pens *Virgularia mirabilis*, do occur in Plymouth Sound but they are much more abundant in some other muddy harbours. The biotope description is based on more northern examples characterised by species that do not occur or are very rarely seen in southern Britain.)

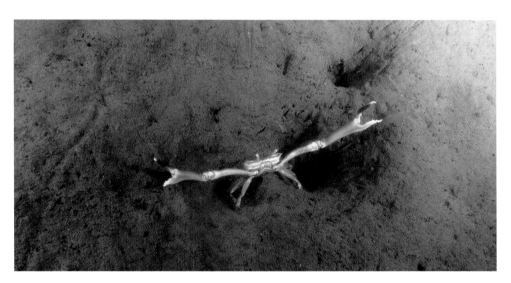

The biotopes that characterise mud habitats are composed of species that live almost entirely below the surface. Burrows may be conspicuous to divers but grabs and dredges rarely penetrate deeply enough to catch burrowing large crustaceans, and fish and the inhabitants of the mud remain unrecorded.

Two species found in mud sediments in shallow depths. The burrowing anemone *Scolanthus callimorphus* (left) and the Slender Sea Pen *Virgularia mirabilis* (right). Both are photographed in Plymouth Sound. Image widths *c.* 7cm.

The echiuran worm *Maxmuelleria lankesteri* is often found in samples from mud in Scottish sea lochs. In the ultra-sheltered conditions of a sea loch, divers may see the volcano-like mounds that this large worm creates at one end of its 'U'-shaped burrow. At the other end, and only in dark conditions, the green proboscis may be seen, but withdraws as soon as it is illuminated with white light. Using red light to locate the creature, Nick Owen has photographed high densities in mud off the south coast of Kent. Image width *c.* 12cm. Image: Nick Owen. A characteristic species of 'Burrowing megafauna and *Maxmuelleria lankesteri* in circalittoral mud' (A5.362 / SS.SMu. CFiMu.MegMax).

Fine sediments are a preferred habitat for the Fan Mussel *Atrina fragilis* – the largest of the bivalve molluscs. Fan Mussels are very rarely seen but a small population was found in Plymouth Sound in 2005 – one of the nine is illustrated here (image width *c.* 20cm). Only the very top protrudes from the sediment and the remaining 30cm of shell is embedded below. The dead shell illustrated was found near to the small live population and is 40cm long. Remarkably, a very dense population was found in the Sound of Canna off the west coast of Scotland in 2010 when more than a hundred Fan Mussels were discovered during a routine Marine Scotland survey.

The Red Bandfish: discovery and life style

The lifestyle of Red Bandfish *Cepola macrophthalma* was a mystery until they were seen by research divers in burrows at Lundy in the early 1970s. They had been caught in trawls and by anglers and it was known that, in aquaria, they sought out pipes and other tube-like structures to live in. But even when the observations at Lundy were reported to specialists in burrowing megafauna, those scientists had to see them for themselves before believing that there was indeed a red fish living in holes. The research that was subsequently undertaken revealed a population of about 14,000 fish off the east coast of Lundy. The fish extended from or left their holes to snap at passing plankton and also had a highly social life, visiting each other's burrows. The results of the work at Lundy were published by Jim Atkinson, Roger Pullin and Frances Dipper in 1977.

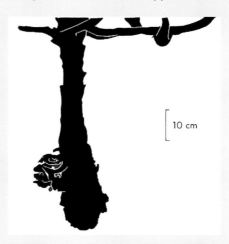

This is the sight that greeted research divers using a towed underwater sledge to survey the muddy gravel sediments off the east coast of Lundy in 1973. But the identity of the fish took a little while to determine. Image width *c.* 10cm.

To discover what the burrows looked like, resin was poured into them (the rather annoyed fish swam out straight away) and, the next day, the casts were carefully excavated.

The burrows were about 50–60cm long with a chamber at the bottom, and the burrows of other species – the Angular Crab *Goneplax rhomboides* and crustaceans such as *Upogebia stellata* – were linked to them. From a paper by Jim Atkinson and Roger Pullin in the Lundy Field Society Annual Report.

Here, in a depth of only 2m below chart datum in Plymouth Sound, a small group of Red Bandfish come out to feed in the water column – snapping at passing plankton on the ebb tide. The fish in the foreground is nervous and doesn't come all the way out – the camera was only about 50cm away. Image width *c.* 80cm in the foreground.

OPEN COAST ROCKY AREAS

Hard stable substrata support the greatest range of seabed communities. They are also the habitats that most attract observation by divers and are the major part of this book. Different communities develop on reefs according to geographical location (biogeography), the physical environment (depth and degree of exposure to wave action and tidal currents especially) and the presence of geomorphological features such as cliffs, canyons and caves. Stable boulders may be colonised on the outer surface by the same species that colonise bedrock. Underneath, however, they offer shelter to species that may live there permanently to avoid the many predators or that come out, perhaps only at night, to forage. The geology of the substratum is also important, with some species favouring soft rocks (chalk or limestone particularly) that can accommodate organisms that bore into the rock (and leave holes for colonisation when they die) and some highly layered rock which forms fissures and ledges. In wave-sheltered locations, the deposition of silt may be important in determining what can grow attached to the rock below. Some reef habitats have already been described in extreme environments or physiographically distinct parts of the coast. In this section, I have included open coast reefs in high and moderate energy environments.

How to organise this section has been a dilemma. In the end, I decided to split Britain north, east and south-east and south-west with examples from different open coast habitats and some snippets that apply everywhere about unusual habitats. But, it is a selection and not comprehensive.

Northern coasts

Shallow subtidal rocks on the coasts of northern Britain (approximately north of Anglesey and of Norfolk) are separated from southern areas by both climate (colder waters) and by a break in seabed type with predominantly sedimentary seabed until Cumbria in the west and Yorkshire in the east. The colder waters mean that some of

On wave-exposed coasts in northern Britain, Dabberlocks *Alaria esculenta*, is a characteristic feature of the area around low water level to a depth of about 1m below chart datum – although that depth extends to 14m on the extremely wave-exposed coasts of St Kilda. Here, it is photographed at Weasel Loch near St Abbs. Image width *c.* 80cm. '*Alaria esculenta* on exposed sublittoral fringe bedrock' (A3.111 / IR.HIR.KFaR.Ala).

the species most likely to be encountered are ones that thrive in colder conditions and some that only occur in northern waters. One such characteristic species, the kelp *Alaria esculenta*, has a peculiar distribution where it is abundant on wave-exposed coasts in the Isles of Scilly and on the offshore reefs at the Eddystone lighthouse but has disappeared from inshore areas of south Devon where it was recorded several decades ago. At Lundy, an intermediate abundance occurs – it is sparse in shallow depths just as it was in the late 1940s when Leslie Harvey described the seashore biology there.

A wave and tide-sheltered gulley at St Abbs in Berwickshire. The forest of Cuvie *Laminaria hyperborea*, dominates the rock and cuts out up to 99% of light. Together with lack of adequate light for photosynthesis, sea urchin grazing keeps the stipes of the kelp clear of growth and the rocky seabed below dominated by calcareous seaweeds with very few foliose algae. Image width *c.* 1.5m. 'Grazed *Laminaria hyperborea* forest with coralline crusts on upper infralittoral rock' (A3.2143 / IR.MIR. KR.Lhyp.GzFt).

In wave-exposed locations where Dead Men's Fingers *Alcyonium digitatum* and Plumose Anemones *Metridium dianthus* thrive, the switch from kelp-dominated rock to rock characterised by these animals may be dramatic, with no intervening zone dominated by foliose algae. Skelly Hole, St Abbs.

Grazing by sea urchins (Edible Sea Urchin *Echinus esculentus* pictured here at St Abb's Head) can greatly reduce the downward extent of the kelp forest and kelp park as well as eliminating any foliose algae that might settle in well-lit waters. Image width *c*. 1.5m. The biotope would have corresponded in the 1997 version of the JNCC classification to '*Echinus*, brittlestars and coralline crusts on grazed lower infralittoral rock' (IR.SIR.K.EchBriCC) but that biotope is now subsumed into 'Grazed *Saccharina latissima* with *Echinus*, brittlestars and coralline crusts on sheltered infralittoral rock' (A3.3134 / IR.LIR.K.Slat.Gz).

An unusual sight in British waters – the Northern Sea Urchin *Strongylocentrotus droebachiensis*, here photographed at Estwick Voe in Shetland and doing what sea urchins do – grazing. Photographed in 1986 but still present at locations in Shetland. Image width *c*. 80cm. 'Echinoderms and crustose communities on circalittoral rock' (A4.21 / CR.MCR. EcCr). Image: Sue Scott/JNCC.

Wherever coarse sediments occur in areas where tidal currents are strong, Dahlia Anemones *Urticina felina* are likely to be present attached to rocks below the gravel. Brittlestars (*Ophiothrix fragilis* here) are likely to form dense beds. Image width *c.* 40cm.

Below left: a typical circalittoral community of open coastal areas sheltered from very strong wave action on the west coast of Scotland (this image is from north of Elian Dubh Mor in the Firth of Lorn). The sea fans here are Northern Sea Fans *Swiftia pallida* and are accompanied by large numbers of the Devonshire Cup Coral *Caryophyllia smithii*, a patch of the White Cluster Anemone *Parazoanthus anguicomus*, several Football Sea Squirts *Diazona violacea* and Dead Men's Fingers *Alcyonium digitatum*. Image width *c.* 80cm. '*Caryophyllia (Caryophyllia) smithii* and *Swiftia pallida* on circalittoral rock' (A4.211 / CR.MCR. EcCr.CarSwi).

Below right: a closer image of the Northern Sea Fan.

Rock surfaces on the open or semi-enclosed coast of northern waters are dominated by kelp *Laminaria hyperborea* in shallow depths and often 'bare' rock that is grazed heavily by Edible Sea Urchins *Echinus esculentus*, deeper than the kelp. The rock is, of course, not bare but colonised by resistant encrusting algae and often has a fuzz of fast-growing algae, sea firs, etc. There may also be dense stands of Dead Men's Fingers *Alcyonium digitatum* and Plumose Anemones *Metridium dianthus*, immediately deeper than the kelp on wave-exposed coasts. Brittlestars *Ophiothrix fragilis* and *Ophiocomina nigra* often thrive here, crawling over the algae-encrusted rocks and the Dead Men's Fingers.

Northern species that are found especially on the north-east coast of England and south-east coast of Scotland (the St Abbs and Farne Islands area).

Above: Northern Tooth Weed *Odonthalia dentata* occurs all around Scotland and to parts of north-west and north-east England. Image width *c.* 12cm.

Above: the Deeplet Sea Anemone *Bolocera tuediae* is almost entirely confined to Scottish waters . Image width *c.* 10cm. Image: Fiona Crouch.

Above: the Bottlebrush Hydroid *Thuiaria thuja* is recorded particularly from south-east Scotland and north-east England. Image width *c.* 10cm. Image: Fiona Crouch.

Right: Atlantic Wolffish *Anarhichas lupus* are found in small numbers on the east coast of Scotland and the Northern Isles. Image width *c.* 14cm. Image: Fiona Crouch.

Whereas the Black Brittlestar *Ophiocomina nigra* is commonly found all around Britain, the Common Sunstar *Crossaster papposus* is only very rarely seen in most south-western waters and abundance seems to have declined there during the 20th century. Sunstars feed on other echinoderms. Image width *c.* 30cm.

Brittlestar beds

Beds of brittlestars occur all around the coast of Britain. The most conspicuous are the three species that occur as dense aggregations on bedrock through to sandy sediments, often crawling over other species. The Common Brittlestar *Ophiothrix fragilis* (shown in the long picture above) is widely distributed and can occur in densities of nearly 2,000 per square metre. Its behaviour was studied in the 1970s by George Warner. When tidal currents are slack or slow, the arms are mainly flat on the seabed but, as the current picks up, the arms are raised into the water to catch suspended food – they are passive suspension feeders and rely on moving water to bring them food. However, when the current strength increases to above about 15cm/s, the brittlestars crouch lower to the substratum and reduce the number of feeding arms. Stronger again (above about 25cm/s) and all the arms are lowered and linked together for stability. Sometimes that stability is not maintained and Swiss rolls of brittlestars have been reported tumbling along the seabed.

The Black Brittlestar *Ophiocomina nigra* can occur as single-species beds or, more often, mixed in with Common Brittlestar *Ophiothrix fragilis* and, in northern waters, the Crevice Brittlestar *Ophiopholis aculeata*. Image width *c.* 20cm.

The Crevice Brittlestar *Ophiopholis aculeata* occurs as part of mixed brittlestar beds in northern waters but seems confined to dwelling in rock crevices in southern waters. Image width *c.* 20cm.

Eastern and south-eastern coasts

Inshore subtidal rocky areas off the east and south-east coast, including Flamborough Head and then from north Norfolk around to the Isle of Wight, will be a mystery to most scuba divers – the water is often turbid – but there are reefs that have rich and unusual communities attached. A particular feature of bedrock along parts of this coast is that much of it is chalk.

Flamborough Head is the most northerly chalk outcrop in Britain. The chalk is fairly hard and extends underwater to shallow depths. The biology of the reefs has been surveyed extensively, especially during the 1980s and 1990s. There are shallow beds of Blue Mussels *Mytilus edulis* and, offshore, reefs of Horse Mussels, *Modiolus modiolus*. Rock communities are colonised by a rich variety of hydroids, sea mats, ascidians and echinoderms, many of which are typical of cold waters. Sea urchin grazing seems much less on chalk substrata and so rich algal communities and turfs of animal species often occur.

Dramatic chalk cliffs near Flamborough Head extend seawards as shallow shelving seabed dominated by mussels, foliose algae and turfs of hydroids and sea mats.

Norfolk's 'Great Barrier Reef'

There are still habitats to be found, explored and described. The rocky seabed off north Norfolk is well known for its crabbing industry but it is only since about 2010 that the extent and structural complexity of chalk reef habitats has been properly recognised. Seasearch surveyors Rob Spray and Dawn Watson have led teams of divers to map and describe the reefs. Over 350 species have been recorded along the 20 miles (32km) of reef and, in 2010, their observations excited the public enough for the finds to be publicised in the national press.

One of the chalk reef escarpments off Sheringham that forms a part of, according to the press, 'Britain's Great Barrier Reef'. The reef here is dominated by foliose algae. Image width c. several metres. 'Communities on soft circalittoral rock' (A4.23 / CR.MCR.SfR) and 'Foliose red seaweeds on exposed lower infralittoral rock' (A3.116 / IR.HIR.KFaR.FoR).

An Edible Crab Cancer pagurus, with mysid shrimps in a scour hole on the chalk reef. Image width c. 25cm.

Highly mobile flint cobbles dominated by encrusting calcareous algae and some sponges adjacent to the solid reef habitats. The cobbles and small boulders provide shelter and protection underneath them for a variety of crustaceans, especially juveniles. Image width c. 25cm.

Chalk reefs occur again along the Kent coast, where there are also sandstone and limestone exposures as well as some areas of clay. The identification guide for divers of Sussex marine life produced by Robert Irving indicates that there are relatively few underwater rocky reefs present off the Sussex coast and that, in shallow depths, they are dominated by foliose algae with a few scattered kelp plants. Under overhangs and in slightly deeper water, sponges, soft corals, hydroids, bryozoans and sea squirts occur. The chalk rocks are colonised by piddocks and by other species that burrow into soft rock (see page 113). There are cliffs and gullies in the chalk bedrock with communities that seem, from the description in Robert Irving's guide, to be very similar to those discovered in recent years off Norfolk.

Southern and south-western coasts

This is a broad area that sweeps from the region of Poole Bay all around to north Wales and often includes spectacular rocky coastlines and equally remarkable underwater landscapes. The rocks are varied from shales to granite and limestone. Along many parts of this coast, the water is often clear and algae extend to depths in excess of 20m below extreme low water level. However, the Bristol Channel east of Lundy has increasingly turbid water along the coast towards Minehead on the English coast and St Donats on the Welsh coast – the furthest east that diving or use of underwater imaging equipment is, from time to time, feasible. The biological character of reef habitats is distinctly Lusitanian, although many species with their centres of distribution well to the south of Britain may be confined to the south-west peninsula of England and to Pembrokeshire in Wales.

The open rocky coasts of south-western areas have been widely studied and photographed, enabling some of the general features of rocky subtidal ecology to be illustrated in this section. The text takes the reader from shallow depths dominated by algae to rocks characterised by animals, giving examples of the different assemblages of species that occur in different conditions of wave action and tidal currents, as well as turbid waters and some more specialised or localised habitats.

From the intertidal, bedrock falls away into the sublittoral fringe around low water level. Here, on the east coast of Lundy, intertidal areas are characterised by Thong Weed *Himanthalia elongata* and foliose red algae. On southern coasts that are not extremely exposed to wave action, it will be Oar Weed *Laminaria digitata* and, deeper, Cuvie *Laminaria hyperborea*, that dominates the sublittoral fringe. Image width *c.* 2m.

Shallow rocks Wherever light is sufficient for photosynthesis, algae have the potential to dominate rocky habitats. The depth to which algae extend therefore depends on the turbidity of the water, but also to modifying factors such as sea urchin grazing and strong wave action, which may allow animals to dominate shallow depths. This is because food is plentiful for them in such conditions.

The canopy of Cuvie *Laminaria hyperborea* overlies an often rich community of species attached to the underlying rock. Here, the kelp is in a moderately sheltered location in the Landing Bay at Lundy and Snakelocks Anemones *Anemonia viridis* can attach to the fronds. Under the kelp, the depths are shallow enough for sufficient light to reach the seabed for a predominantly foliose algal covering of rocks. Image width *c.* 1m. '*Laminaria hyperborea* forest and foliose red seaweeds on moderately exposed upper infralittoral rock' (A3.2141 / IR.MIR. KR.Lhyp.Ft).

Above: shallow open coast unstable rocky habitats, wherever they occur in Britain, are characterised by a dense cover of ephemeral algae during summer. Here, photographed at Porthkerris on the Lizard Peninsula. Image width *c.* 1m. '*Saccharina latissima*, *Chorda filum* and dense red seaweeds on shallow unstable infralittoral boulders and cobbles' (A3.123 / IR.HIR.KSed. SlatChoR).

Left: in wave-sheltered areas (this is at West Hoe in Plymouth Sound) in the south-west, warm water Golden Kelp *Laminaria ochroleuca*, has become abundant and taken over from Cuvie *Laminaria hyperborea* in places. Image width *c.* 2m. 'Mixed *Laminaria hyperborea* and *Laminaria ochroleuca* forest on moderately exposed or sheltered infralittoral rock' (A3.311 / IR.LIR.K.LhypLoch).

Under the kelp (*Laminaria hyperborea* and *Saccorhiza polyschides*) on wave-exposed coasts with seaweeds (mainly *Dilsea carnosa*) and cushion sponges (*Amphilectus fucorum*) on the pinnacle of Hilsea Point Rock east of Plymouth and at a depth of about 8m. Image width *c.* 80cm. 'Kelp with cushion fauna and/or foliose red seaweeds' (A3.11 / IR.HIR.KfaR).

Lower infralittoral rocks dominated by foliose algae at White Beach on the east coast of Anglesey. Here, sea urchins are sparse and so grazing pressure is low. The habitat is only about 0.5m below chart datum (extreme low water of spring tides) and lack of a *Laminaria hyperborea* kelp forest is a reflection of the normally high turbidity of the water. Algae are mainly Cartilaginous Cock's Comb *Plocamium cartilagineum*, Red Rags *Dilsea carnosa* and young kelp *Laminaria* sp. plants. Image width *c.* 80cm. 'Foliose red seaweeds on exposed lower infralittoral rock' (A3.116 / IR.HIR.KfaR.FoR) or '*Laminaria hyperborea* park and foliose red seaweeds on moderately exposed lower infralittoral rock' (A3.2142 / IR.MIR.KR.Lhyp.Pk).

Deeper rocks Deeper than the zone of algal domination, rocks are colonised by an often wide variety of animal species. Differences in the species and communities present are brought about mainly by the degree of exposure to wave action and the strength of tidal currents. Some of those animal species will be very conspicuous and colourful, with even hardened wreck divers commenting on their beauty. In south-western waters, it seems that grazing by sea urchins is not as severe as in many northern waters and sparsely colonised or bare rocks are rarely seen unless sand-scoured. There are species that may thrive in certain habitats such as sand-covered rocks or where suspended sediment levels are high.

The Pink Sea Fan and Ross 'coral' biotope is one of the most widely distributed and common biotopes on circalittoral rock in south-west England. Here, offshore of Plymouth Sound. Image width *c.* 60cm '*Eunicella verrucosa* and *Pentapora foliacea* on wave-exposed circalittoral rock' (A4.1311 / CR.HCR.XFa.ByErSp.Eun).

The iconic species of south-west reefs: Pink Sea Fans, *Eunicella verrucosa* (also in white) in Bigbury Bay, south Devon. The white variety is rare off mainland coasts but more abundant in the Isles of Scilly. Moving further south along the European coasts, about 20% are white in Brittany through to 100% off Gibraltar. Image width *c.* 30cm.

These sponge gardens are also a part of the *Eunicella verrucosa – Pentapora foliacea* biotope. The branching yellow species is Yellow Staghorn Sponge *Axinella dissimilis* which has been monitored for growth rates that were found to be less than 1mm a year. Blackstone Point, south Devon. Image width *c.* 80cm.

The top of a cliff at 25m depth that falls almost vertically to 45m at Hatt Rock offshore of Looe in Cornwall. The rock is dominated by Jewel Anemones *Corynactis viridis*, typical of such offshore reefs exposed to strong wave action and moderate tidal currents. Image width (foreground) *c.* 2m. '*Corynactis viridis* and a mixed turf of crisiids, *Bugula*, *Scrupocellaria*, and *Cellaria* on moderately tide-swept exposed circalittoral rock' (A4.132 / CR.HCR. XFa.CvirCri).

Jewel Anemones *Corynactis viridis* put on a spectacular display, especially where tidal currents and/ or wave action are strong. Image width *c.* 4cm.

The most conspicuous and characteristic species of fish associated with reef habitats are wrasse. They are particularly a feature of southern coasts and become less abundant in the north of Britain.

A Ballan Wrasse *Labrus bergylta*. These are the largest of the wrasse and mainly inhabit shallow waters, where they forage for food amongst algae or the turf of hydroids and bryozoans. Image width *c.* 30cm.

A male Corkwing Wrasse *Symphodus melops*. The females are less colourful. Males hold territory and fiercely defend the nests that they build in mainly shallow depths. Image width *c.* 16cm.

A female and a male phase Cuckoo Wrasse *Labrus mixtus*. The female is only slightly smaller than the male but perspective in this image makes the female look much smaller. All Cuckoo Wrasse start life as females but some will dramatically change colour and become fully functional males. They are seen mainly deeper than the kelp forest. Image width in the foreground *c.* 35cm.

Goldsinny Wrasse *Ctenolabrus rupestris* are present in broken rocky areas deeper than the kelp forest. They occur mainly on wave-sheltered coasts including in rias and the outer parts of estuaries. Image width *c.* 12cm.

Shoals of Rock Cook *Centrolabrus exoletus* are most common amongst the kelp forest in open coast areas and are frequently seen accompanying Ballan Wrasse picking up food scraps and cleaning the wrasse of parasites. Image width *c.* 10cm.

Revealing the secrets of Black Bream breeding behaviour on the rocky seabed off the Dorset coast

A male Black Bream guards an area of rock that has been cleared of sediment to provide a nest where eggs can be laid. Males have a white vertical stripe on their side.

The seabed, where up to 70kg of sediment has been cleared to create each nest, can look like craters on the moon.

There remains much to learn about the ecology of subtidal rocky seabeds and their inhabitants. Each spring, hundreds of thousands of Black Bream *Spondyliosoma cantharus* arrive along the coast of southern England and west Wales to breed. Divers usually see only skittish individuals and the areas of rock cleared of sediment ready for eggs to be laid. It took patience and cunning to film them. By placing video and time-lapse cameras on the seabed and then retreating, Matt Doggett and colleagues have captured stunning images of the fish and documented their breeding behaviour. When the eggs had been laid, it could be seen that they were a rich source of food for other fish and for some invertebrates – if they could get past the aggressive guarding of the males. Images: Matt Doggett.

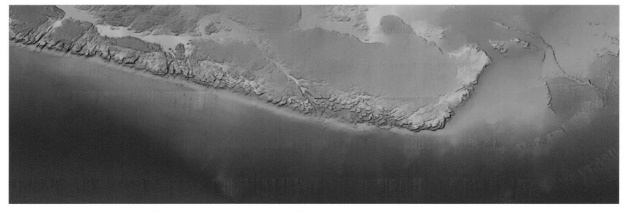

A structurally complex area of reef known as The Dropoff, offshore of Plymouth Sound imaged by multi-beam sonar. The red-coloured area is about 27m in depth and the blue-coloured area of level seabed is about 38m depth below chart datum. The cliffs and gullies provide habitats favoured by a wide range of rare or scarce (and often colourful) species. The length reef in the image is about 1km. Image courtesy of the School of Biological and Marine Sciences, University of Plymouth.

Vertical drops and gullies on a submerged cliffline off Plymouth at 30–40m depth. Image width *c.* 1m.

Cliffs colonised by species mainly with a southern or western distribution, including some illustrated separately opposite. Image width *c.* 50cm. 'Sponges, cup corals and anthozoans on shaded or overhanging circalittoral rock' (A4.711 / CR.FCR.Cv.SpCup).

Some reef species with a predominantly south-west and west coast distribution:

Left: Golden Kelp *Laminaria ochroleuca* was first reported from British waters in 1948 and is now frequently seen, especially in wave-sheltered areas but only in south-west England. Image width *c.* 2m.

Right: Rainbow Wrack *Cystoseira tamariscifolia* is a shallow-water alga that occurs off south-west and western coasts. Image width *c.* 30cm.

Left: the Yellow Cluster Anemone *Parazoanthus axinellae* has a south-west and west coast distribution, occurring as far north as the Outer Hebrides and St Kilda. Here, it is typically growing on the sponge *Axinella damicornis*, which also has a south-west and west coast distribution. Image width *c.* 12cm.

Right: the Sunset Cup Coral *Leptopsammia pruvoti* is known from six locations in south-west England and has its centre of distribution in the Mediterranean. Image width *c.* 4cm.

Left: the Trumpet Anemone *Aiptasia couchii* is recorded in south-west Britain and eastwards into Poole Bay. Image width *c.* 10cm.

Right: Red Sea Fingers *Alcyonium glomeratum* is a soft coral with a south-western and west coast distribution extending to the Outer Hebrides and St Kilda. Image width *c.* 20cm.

Left: the Spiny Spider Crab *Maja brachydactyla* is a southern and west coast species. There are records on the east coast but the species is easily confused with the Northern Stone Crab *Lithodes maja*. Image width *c.* 30cm.

Right: the orange sea squirt *Stolonica socialis* is a south-west and west coast species with only a very few records from Scotland. Image width *c.* 7cm.

Not all reefs are of stable substrata. Many are composed of pebbles, cobbles and boulders that may be moved by storms. Such rock habitats have a covering of fast-growing algae in shallow depths and, deeper, of hydroids, sea mats and tube worms. Some of those species are characteristic of mobile substrata. Rocks near to coarse sediment may also become covered in a layer of sand or gravel as well as suffering scour during storms. A different suite of species characterise those sand-covered or scoured reef habitats.

Sand-covered rocks and seasonally mobile cobbles and pebbles Some perennial algae can survive at least their holdfasts and lower stipes being inundated by sand. Others are annual species that are characteristic of both sandy rocks and of seasonally mobile cobbles and pebbles.

Shelter from strong wave action but with moderate tidal currents creates an environment in which the richest communities of sessile species occur. Here, a community off the east side of St Mary's in the Isles of Scilly is shown. Image width c. 1m. '*Eunicella verrucosa* and *Pentapora foliacea* on wave-exposed circalittoral rock'. (A4.1311 / CR.HCR.XFa.ByErSp.Eun).

Some algae typical of sand-covered rock or seasonally unstable cobbles and pebbles:

The shifting sands of a tidal sound – here Smith Sound in the Isles of Scilly between Annet and St Agnes. The plants are attached to stable cobbles and pebbles but have to withstand cover by sand. Cuvie *Laminaria hyperborea* is in the background and Pod Weed *Halidrys siliquosa* in the foreground. The image is taken at about 15m below chart datum. Image width *c.* 2.5m. The biotope is difficult to classify but should fall within 'Infralittoral coarse sediment' (A5.13 / SS.SCS. ICS). [More likely one of A3.12]

Left: *Furcellaria lumbricalis* in Smith Sound, Isles of Scilly. Image width *c.* 40cm.

Right: Papery Fan Weed *Stenogramma interruptum* (with the hydroid *Tubularia indivisa*) in Smith Sound, Isles of Scilly. Image width *c.* 25cm.

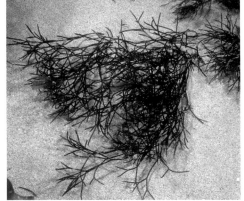

Left: Scinà's weed *Scinaia* sp. south of Plymouth Sound breakwater. Image width *c.* 10cm.

Right: Hand Leaf Bearer *Phyllophora sicula* off Stoke Point, south Devon. Image width *c.* 8cm.

Cobbles, pebbles and small boulders at wave-exposed infralittoral locations are likely to be turned over during strong wave action and attract a community characterised by ephemeral algae and coralline crusts. Species such as the Snakelocks Anemone *Anemonia viridis* are loosely attached and may re-settle. Here, a photograph taken south of Plymouth Sound breakwater. Image width *c.* 50cm. Most likely one of the biotopes that are part of 'Sediment-affected or disturbed kelp and seaweed communities' (A3.12 / IR.HIR.KSed).

Tide-swept cobbles and pebbles on sand are colonised mainly by ephemeral foliose algae and more robust encrusting calcareous algae that survive being turned over and abraded during winter storms. Photographed at Smith Sound, Isles of Scilly. Image width *c.* 1m. The biotope could only be identified to 'Sublittoral coarse sediment' (A5.1 / SS.SCS).

The species colonising this upper circalittoral shell-gravel-covered rock need to tolerate potential smothering and scour and some may especially favour scoured conditions. Here, the flat-fronded red alga is *Stenogramma interruptum*, which is characteristic of seasonally disturbed cobbles and pebbles and of sand-abraded rocks. Other species include the Yellow Hedgehog Sponge *Polymastia boletiformis*, a Pink Sea Fan *Eunicella verrucosa* and Dead Men's Fingers *Alcyonium digitatum*. Photographed at the Eddystone reefs. Image width *c.* 40cm. The biotope is most likely a part of 'Bryozoan turf and erect sponges on tide-swept circalittoral rock' (A4.131 / CR.HCR.XFa.ByErSp) but probably needs a separate biotope to accommodate a community characteristic of sandy rocks.

Animal species that are characteristic of sandy bedrock are mainly sponges and anemones. Several are also found on rock covered by silt. Some that typically occur in or near coarse sediments are shown below.

Left: the Antler Sponge *Adreus fascicularis* is nationally scarce and occurs on sand-covered rocks . Image width *c.* 20cm.

Right: the Cored Chimney Sponge *Ciocalypta penicillus* is highly characteristic in appearance. It has a basal cushion that attaches to rock below gravel and sand. Image width *c.* 8cm.

Left: Dahlia Anemones *Urticina felina* are widespread but favour rocks with a layer of gravel. Image width *c.* 10cm.

Right: the Peppercorn Anemone (also known as Ginger Tinies) *Isozoanthus sulcatus* grows from stolons that meander over bedrock below the sediment cover. Image width *c.* 15mm.

Sand-covered rocks at Stoke Point in south Devon colonised by, in the foreground, the sponge *Adreus fascicularis* and, on the distant part of the rock outcrop, *Ciocalypta penicillus*: both characteristically found on sand-covered bedrock. Image width *c.* 80cm in the foreground. The biotope is most likely a part of 'Bryozoan turf and erect sponges on tide-swept circalittoral rock' (A4.131 / CR.HCR. XFa.ByErSp) but likely needs a separate biotope to accommodate a community characteristic of sandy rocks.

Turbid waters There are regional characteristics to the marine life around British coasts that are not associated with seawater temperature and the geographical distribution of species, and it seems that turbidity (sediment loading) of the water may be important. An area that I am familiar with includes the mainland coast of north Devon and its offshore island, Lundy. Turbidity of the water increases from Lundy eastwards towards Bristol. Diving along the north Devon coast has plenty of 'good moments' for visibility but the furthest east I have managed to dive in the Bristol Channel is off Minehead in Somerset.

The depth to which seaweeds can grow on reefs is greatly restricted by poor light penetration, and the downward extent of the kelp forest decreases from 8m below extreme low water level at Lundy to scarcely a metre between Combe Martin and Lynmouth. Deeper than seaweeds, reefs are especially dominated by branching sea mats – a rich habitat for associated species – and sponges. There may be mussel beds but, also, areas that are scoured by coarse sediments that are highly mobile in the strong currents that occur off north Devon. The level seabed offshore of north Devon is largely composed of those coarse mobile sediments with a very few ephemeral attached species.

The turf of branching sea mats (bryozoans) and hydroids at Lundy contains a rich variety of associated species. In this area off the west coast, over 200 species have been recorded from observations and from samples. It is the turf that holds the largest number of species, including tiny crustaceans, snails and polychaete worms. What the diver sees or the video camera records are the large organisms such as Hornwrack *Flustra foliacea* (at the back of the image), the Pink Sea Fan *Eunicella verrucosa* and Ross 'coral' *Pentapora foliacea*. Image width *c.* 50cm. '*Eunicella verrucosa* and *Pentapora foliacea* on wave-exposed circalittoral rock' (A4.1311 / CR.HCR.XFa.ByErSp.Eun).

Similar communities to the ones at Lundy on stable boulders on the nearby north Devon coast off Baggy Point. Typically, there are more antenna hydroids *Nemertesia antennina*, and Sea Chervil *Alcyonidium diaphanum* than at Lundy and, although they do occur on the north Devon mainland coast, no Pink Sea Fans. Image width *c.* 1.5m. 'Very tide-swept faunal communities on circalittoral rock' (A4.11 / CR.HCR.FaT).

Cobbles and small boulders are turned over by wave action and are sparsely colonised by fast-growing ephemeral species. The larger boulders are more stable and have the greatest amount of growth. Dahlia Anemones *Urticina felina*, are characteristic of such areas. The Common Starfish *Asterias rubens* feed on barnacles *Balanus crenatus*. Off Baggy Point, north Devon. Image width *c.* 80cm. Most likely part of 'Mixed faunal turf communities on circalittoral rock' (A4.13 / CR.HCR.XFa) but the scoured nature of the small boulders needs a specific biotope.

The north-facing aspect of the coastline of north Devon protects the seabed from prevailing winds, and strong wave action is only occasional. Turbid waters mean that rich, animal-dominated communities occur in shallow waters. Here, cushion and branching sponges are seen amongst a silty turf of branching sea mats in Combe Martin Bay at a depth of about 12m below chart datum. Image width *c.* 24cm. 'Cushion sponges and hydroids on turbid tide-swept sheltered circalittoral rock' (A4.251 / CR.MCR. CFaVS.CuSpH) or 'Bryozoan turf and erect sponges on tide-swept circalittoral rock' (A4.131 / CR.HCR. XFa.ByErSp).

Reefs are attractive to shoaling fish. Here Bib *Trisopterus luscus* are pictured at Copperas Rock east of Combe Martin in north Devon. Image width *c.* 70cm.

Subtidal mussel (*Mytilus* spp.) beds In situations where large mussel beds are present on intertidal rocks and, seemingly, where there is often sand in suspension, mussel beds may also develop offshore – such mussel beds are a feature of north Cornwall. Mussel beds may also develop in turbid waters and may be ephemeral: often consumed by starfish and then absent for many years. They also develop in areas exposed to very strong tidal currents, such as the bed illustrated at Horseshoe Rocks in north Devon in the section on 'Tidal sounds and other extremely tide-swept habitats' (page 169), at Portland Bill, and other extremely tide-swept habitats.

A rock pinnacle dominated by Blue Mussels *Mytilus edulis*, with Spiny Spider Crabs *Maja brachydactyla* between Westward Ho! and Clovelly. The adjacent rock pinnacle was dominated by foliose algae, demonstrating the patchy occurrence of some species under apparently very similar conditions. Location: Greencliff Rock at about chart datum level. Image width *c.* 1.5m. (It has not been possible to match the mussel dominated community to a higher level biotope than 'Atlantic and Mediterranean moderate energy infralittoral rock' (A3.2 / IR.MIR).)

Caves, tunnels and overhangs Volcanic action is rather rare in Britain, which may be one reason why we have so few underwater caves. And, if there are ancient submarine river exits from limestone rocks, they have yet to be found. There are caves in the intertidal zone – usually where softer rock has been eroded by wave action and that wave action continues to scour the sides with sand, pebbles and cobbles, leaving only impoverished communities. Those intertidal caves may occasionally extend into the subtidal: for instance at Staffa in the Inner Hebrides, at Eshaness in Shetland and at Berry Head in south Devon. The sea cave-diving holy grail of British scuba diving must be St Kilda but, not having dived there despite two attempts, mine must be a vicarious understanding seen through the videos and still photographs of others. Underwater, in more accessible diving locations, there are narrow gullies, tunnels and overhangs, but not the sort of elongated or dead-end fully submerged caves that attract distinct communities and even include endemic species in some parts of the world. Nevertheless, the cave-like habitats in British waters are populated by certain species that are not often found on more open rock surfaces. Some of the occupants that characterise those cave-like habitats seem to thrive because of shade, some because of strong wave action, and some because of both. The biotopes classification

A diver emerges from one of the tunnels on the coast to the east of Plymouth. Here, at Hilsea Point Rock.

separates overhangs and caves into shallow water examples that are 'Robust faunal cushions and crusts in surge gullies and caves' and, deeper, 'Communities of circalittoral caves and overhangs'.

Shallow surge gullies, caves and tunnels Shallow caves and cave-like habitats are likely to occur in wave-exposed locations where water movement is strong and there is a plentiful supply of food for animals that are passive suspension feeders and opportunistic carnivores.

Left: encrusting sponges *Myxilla* sp., White Clathrina *Clathrina coriacea*, sea squirts *Distomus variolosus*, hydroids with the foraminiferan *Halyphysema tumanowiczii* attached and pale-coloured Elephant-hide Sponge *Pachymatisma johnstonia*. Image width *c.* 10cm.

Right: nearby, tough calcareous sponges are a feature of shallow wave-surged tunnels. Here, *Leuconia nivea* and White Clathrina *Clathrina coriacea*, an encrusting sea mat and Jewel Anemones *Corynactis viridis*. Image width *c.* 10cm.

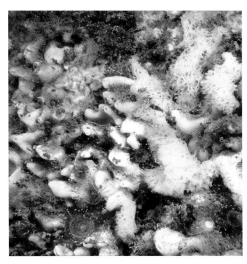

Shallow tunnels without significant scour may be dominated by encrusting sponges with sea squirts, hydroids and Jewel Anemones. Here at about 5m depth at Long Ray, east of Hilsea Point, east of Plymouth. Image widths *c.* 10cm. 'Crustose sponges on extremely wave-surged infralittoral cave or gully walls' (A3.715 / IR.FIR.SG.CrSp).

At the ends of shallow tunnels and gullies that funnel wave action, a community of Baked Bean Ascidian *Dendrodoa grossularia* and White Clathrina *Clathrina coriacea* typical of wave surge conditions occurs. Here, at Men-a-vaur in the Isles of Scilly, the main characterising species are joined by Jewel Anemones *Corynactis viridis*. Notably, the sea squirt seems to characterise habitats at two ends of the wave exposure spectrum – in extreme surge conditions and in marine inlets where it is unaffected by wave action, although tidal currents may be strong. Image width *c.* 8cm. '*Dendrodoa grossularia* and *Clathrina coriacea* on wave-surged vertical infralittoral rock' (A3.714 / IR.FIR.SG.DenCcor).

The Scarlet-and-gold Star Coral *Balanophyllia regia* is frequently found on the lower parts of shallow-water caves and tunnels. Here, with the Devonshire Cup Coral *Caryophyllia smithii*, at Porthkerris on the Lizard Peninsula, in a small underwater cavern known as the 'Crack of Life'. Image width *c.* 6cm.

Overhangs Shaded rocks that are not subject to a great deal of water movement (but are well oxygenated and there is sufficient water movement to bring food) provide a benign habitat for species that are intolerant of well-lit conditions. They are especially attractive to encrusting sponges, to branching sea mats (which have a rich associated fauna of mobile species that may provide food for opportunistic carnivores) and to soft and stony corals.

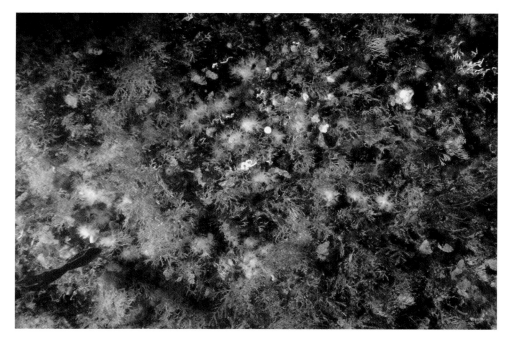

Caves and shaded overhangs sometimes have colonies of nationally scarce Pink Sea Fingers *Alcyonium hibernicum*. The colonies are not widespread and occur in localised areas. Here, at Martin's Haven, the colony was first found on 8 March 1975 and was re-located in exactly the same place on 5 September 2012. It remains in the same place more than 40 years after the first record. Image width *c.* 18cm. 'Sponges, cup corals and anthozoans on shaded or overhanging circalittoral rock' (A4.711 / CR.FCR.Cv.SpCup).

Life under the ledge

Underwater canyons, such as this one at about 15m depth off the east coast of Lundy, are likely to include shaded overhangs that have cave-like features. Overhangs need close inspection with a torch to discover many of the special features. There may also be (but not at Lundy) Black-faced Blennies *Tripterygion delaisi*, often upside down under the overhang. In this location, four of the five shallow-water species of stony corals found in British waters can be seen on the overhanging surfaces – the fifth, the Scarlet-and-gold Star Coral *Balanophyllia regia*, is found in shallower depths nearby. At Lundy, the abundance of some of the southern species has declined since the mid-1980s, possibly the outcome of decadal-scale changes including poor recruitment to replace lost individuals. 'Sponges, cup corals and anthozoans on shaded or overhanging circalittoral rock' (A4.711 / CR.FCR.Cv.SpCup).

Overhanging rock at the Knoll Pins on Lundy with scattered Sunset Cup Corals, *Leptopsammia pruvoti*. Image width *c.* 1m.

Devonshire Cup Coral *Caryophyllia smithii*. Image width *c.* 3cm.

Southern Cup Coral *Caryophyllia inornata*. Image width *c.* 3cm.

Sunset Cup Coral *Leptopsammia pruvoti*. Image width *c.* 4cm.

Weymouth Carpet Coral *Hoplangia durotrix*. Image width *c.* 2cm.

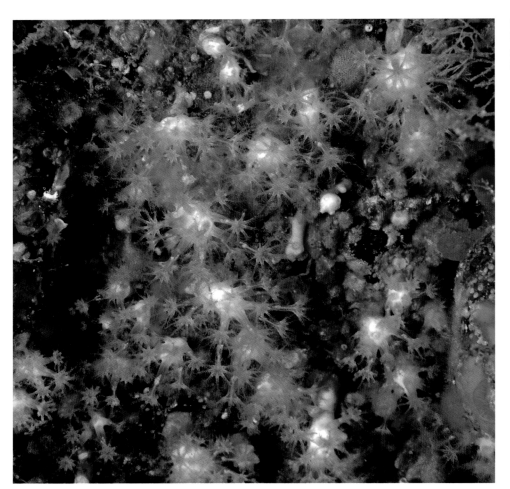

Pink Sea Fingers *Alcyonium hibernicum* on overhanging rocks at Porthkerris on the Lizard Peninsula. Image width *c.* 5cm.

Some mobile species seek shelter under overhangs. Here, a Black-faced Blenny *Tripterygion delaisi* in a small cave at Porthkerris and, typically, upside down. The species was first recorded in Britain near Weymouth in 1972, in the region of Plymouth in 2004 and in the region of Falmouth by 2011. Image width *c.* 12cm.

Living 'in' the rock Most bedrock reefs will have holes, fissures and crevices that offer shelter to species that can squeeze into those narrow spaces. Some are attached to the underlying substratum and just the feeding structures (tentacles or siphons) protrude. Many mobile species seek refuge in crevices, fissures and small caves during the day at least. Some use the holes, fissures, and crevices to lay eggs.

Left: the tubes of the Double Spiral Worm *Bispira volutacornis* are in rock crevices where the worms are protected from predators. The tube with its characteristic closure is on the right. Image width *c.* 8cm.

Right: a recently recruited European Spiny Lobster *Palinurus elephas* needs to be out of reach of predators and therefore find a hole to live in. As they grow larger, they are found on rock shelves and are largely unprotected by the reef. Image width *c.* 10cm.

Left: the Spiny Squat Lobster *Galathea strigosa* is only ever seen in or at the entrance to a crevice or under boulders, and relies on the cryptic habitat for safety at least during the day. Image width *c.* 8cm.

Right: the Green Sea Urchin *Psammechinus miliaris* is prey to wrasse during daytime and so, in southern Britain, hides under boulders or in rock fissures. Image width *c.* 8cm.

The Tompot Blenny *Parablennius gattorugine* shelters in holes and crevices and under boulders when not out foraging. The female lays eggs within a crevice and the eggs are then guarded by the male occupant. Image width *c.* 7cm.

These female Small-spotted Catsharks *Scyliorhinus canicula* crowd into holes in the reef during the day to hide from the unwanted attentions of males. Image width *c.* 30cm.

Even large predatory species such as this Conger Eel *Conger conger* will find boulder-holes or small caves to shelter in when not foraging. This one is, as is often the case, accompanied by Common Prawns *Palaemon serratus* which also live in fissures and under overhangs. Image width *c.* 18cm.

Some rock habitats are soft enough for specialist borers to create or open up holes and expand crevices to nestle into. Species such as the boring sponge *Cliona* sp., the polychaete worms *Polydora* sp., and the Horseshoe Worm *Phoronis hippocrepia* most likely use acid to penetrate the rock. In the case of bivalve molluscs, creation of a burrow is achieved by scraping with the shell to enlarge a crevice or hole, although chemicals may be involved. The biology of species that bore into rock and wood was a special interest of C. M. Yonge who wrote *The Sea Shore*, published in 1949.

Examples of species that burrow into soft rock such as chalk and limestone:

Sieve-like inhalant and exhalant siphons of the sponge *Cliona* sp. The majority of tissue is in the limestone rock. Image width *c*. 5cm.

The Horseshoe Worm *Phoronis hippocrepia* lives in soft rocks and in crusts of calcified algae on harder rocks. Image width *c*. 3cm.

Tubes of the polychaete worm *Polydora* sp. living in limestone rock. Image width *c*. 2cm.

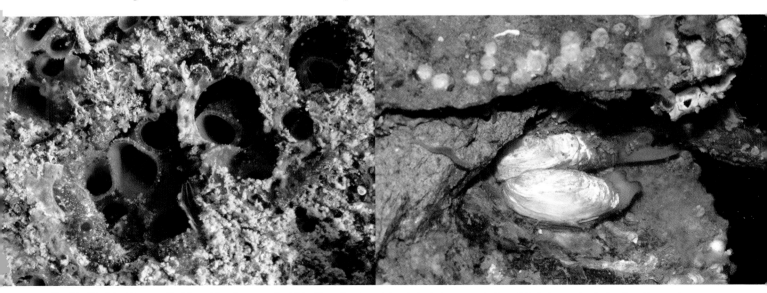

The Wrinkled Rock Borer or Red Nose *Hiatella rugosa* (a piddock) can live in rock in a hole that was created most likely by mechanical action. Here, a view from the surface of the respiratory and feeding siphons (Image width *c*. 3cm), and a view of two where the surrounding rock has fallen away (Image width *c*. 5cm).

WRECKS AND OTHER ARTIFICIAL SUBSTRATA

Humans have been building or losing artificial structures in the sea for millennia. Breakwaters, including many harbours and causeways, create often structurally complex habitats of quarried stone that may be colonised by communities that look natural (but may lack some long-lived, slow-growing species that settle very infrequently, such as some of the branching sponges). Harbours and ports need wharves that are flat-sided and with wooden fenders that may not have structural complexity unless built of blocks that have fissures and crevices between them. Jetties have vertical pilings and cross-members, usually of steel, that are often in strong tidal currents and are shaded below the walkways and roads above, providing a habitat for an abundance of species. Oil rigs, wind farm pilings and other offshore structures are mostly made of steel and are colonised by distinctive communities dominated by a few conspicuous species that often support a rich variety of smaller organisms, as well as attracting large shoals of fish. Such structures may also be protected from scour by boulders placed around their base. Marina pontoon floats are an unusual seabed habitat but attract distinctive subtidal communities. Perhaps the artificial substrata most studied for their biological communities are wrecks. Tens of thousands of ships have been lost at sea around Great Britain. Wrecks are greatly attractive to recreational divers and the photographs they take may be a rich source of information about the larger species that colonise wrecks. More wrecks are now being discovered, as hydrographic surveys of the seabed using multi-beam and sidescan sonar explore areas previously surveyed using only crude techniques.

This section looks especially at the communities that develop on wrecks and the species that favour wrecks as a habitat. Most of the wrecks that a diver will encounter (and that are included here) qualify as the biotope '*Alcyonium digitatum* and *Metridium dianthus* on moderately wave-exposed circalittoral steel wrecks' (A4.721 / CR.FCR. FouFa.AdigMdia). However, some that are in strong tidal currents and that are dominated by the hydroid *Tubularia indivisa*, may be closer to the biotope '*Tubularia indivisa* on tide-swept circalittoral rock' (A4.112 / CR.HCR.FaT.CTub). Breakwaters and reefs made of concrete blocks or quarried stone will most likely match natural reef biotope descriptions. Images in this section therefore do not have biotope names and codes separately.

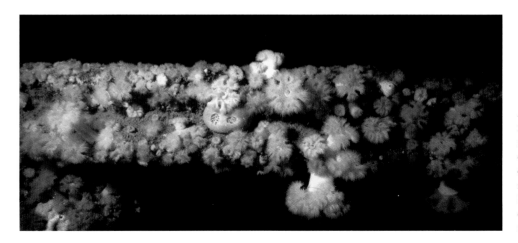

A spar on the wreck of the SS *Hispania* in the Sound of Mull near Oban in August 2004 – it could equally be a cross member on jetty piles or on offshore structures. The spar is dominated by Plumose Anemones *Metridium dianthus* with a sponge *Suberites* sp. Image width *c.* 1m.

Wrecks

Typically, wrecks are of steel but, during the Second World War, concrete harbour sections were constructed for towing to France to create jetties for the invasion and afterwards. Some of them sank before reaching the Channel ports or their eventual destination. These sections of Mulberry harbours now form significant structures at shallow depths off the coast of Sussex, and one lies in deep water west of Penzance. Concrete should provide a surface for attachment that is closer in character to natural rock reefs than steel. The Penzance Mulberry has very similar attached and associated species to those seen on nearby rocks at a similar depth, although it lacks some of the branching sponges and such species as Yellow Cluster Anemones *Parazoanthus axinellae*. These species may be long-lived and slow growing, recruit infrequently and/or have short-lived larvae that have failed to reach hard substrata surrounded by sediment.

Shallow parts of wrecks in areas where the water is reasonably clear are dominated by algae. Here, at about 6m below chart datum, is the bow of the *James Eagan Layne* in Whitsand Bay west of Plymouth, photographed in August 2009. The algal-dominated biotopes will most likely correspond to bedrock reef biotopes.

Deeper, here on the propeller of the *Glanmire* at about 30m depth off St Abbs in May 2014, animals dominate and there are few or no seaweeds. The main species are Dead Men's Fingers *Alcyonium digitatum* (brown and white varieties) with a Common Sunstar *Crossaster papposus*. Image width *c.* 1m.

Collapsed parts of wrecks, where water movement is slight, are often sparsely colonised, although they provide refuges for fish and crustaceans. The propeller shaft and associated plates on the wreck of the *Persier* is pictured in Bigbury Bay east of Plymouth in July 2014, after the severe winter storms of 2013/14 had turned over much of the wreck. Image width *c.* 1.5m in the foreground.

Below: exploring wrecks is a favourite activity for recreational divers and their photographs can contribute to our understanding of what settles readily on new structures. A diver from Plymouth Sound Dive Club inspects part of the steering quadrant on the wreck of the *Persier* in Bigbury Bay in March 2013. Tidal currents are mostly weak here and the most conspicuous species is Dead Men's Fingers *Alcyonium digitatum*.

The parts of wrecks that stand clear of the seabed and extend out into water currents have the richest communities. Here, part of the *Volnay* off the Lizard Peninsula in Cornwall where tidal currents are fairly strong and Plumose Anemones *Metridium dianthus* the most conspicuous species, photographed in June 2013.

Below: in locations exposed to strong tidal currents, such as the bowsprit of the *Maine* off Bolt Tail near Salcombe, shown here, protruding wreck surfaces are most likely to be dominated by tubular hydroids *Tubularia indivisa*. This image was taken in mid-April 2009 but, a few weeks later, the hydroid polyps would probably have been consumed by sea slugs. Image width *c.* 80cm.

The upward-facing plates of wrecks off the south coasts of Devon and Cornwall may be colonised by forests of Pink Sea Fans *Eunicella verrucosa*. Pictured: the *Persier* in Bigbury Bay west of Plymouth in March 2011. Image width *c.* 80cm.

Below: some wreck surfaces can look like natural rock communities. Here, on the side of HMS *Elk* off Plymouth, Red Sea Fingers *Alcyonium glomeratum*, a Pink Sea Fan *Eunicella verrucosa*, Jewel Anemones *Corynactis viridis* and Football Sea Squirts *Diazona violacea* are notable. The image was taken in July 2016, after the great increase in abundance of Football Sea Squirts post 2008. Image width *c.* 1m.

The boiler tubes on the wreck of a steam-powered vessel provide an apartment block for species seeking shelter. The most conspicuous in this image, from the *Persier* in July 2011, is the tube worm *Protula tubularia* with its red tentacles. Image width *c*. 40cm.

Two images from the large, hollow square-section concrete structure offshore of Lamorna Cove west of Penzance in August 2008. This wreck lies on the seabed at about 50m depth. It is 70m long and 17m both high and wide. The structure was a Mulberry, part of a harbour, most likely being towed to a Channel port prior to D-Day in 1944 when it sank.

Horizontal surface at 31m below chart datum characterised by Pink Sea Fans *Eunicella verrucosa*, Dead Men's Fingers *Alcyonium digitatum*, brittlestars *Ophiocomina nigra* and branching sea mats. Corals *Caryophyllia smithii*, were also abundant and there were patches of Rosy Featherstars *Antedon bifida* amongst much else. Image width *c.* 60cm in the middle.

Aperture with edges characterised by species typical of where water movement is accelerated, including Plumose Anemones *Metridium dianthus*, Dead Men's Fingers *Alcyonium digitatum*, Elegant Anemones *Sagartia elegans* and the sponge *Haliclona viscosa*. The Boring Sponge *Cliona celata* is also present. Image width *c.* 80cm.

Some species (other than Plumose Anemones and Dead Men's Fingers in the images above and below) that favour or may be especially abundant on or inside wreck habitats:

A John Dory *Zeus faber* on the wreck of the *James Eagan Layne*. The John Dory is particularly attracted to wrecks, most likely because there is a plentiful supply of food in the form of small fish. Image width *c.* 30cm.

Conger Eels *Conger conger* can shelter under steel plates, in boilers and in exhaust pipes. Here, a Conger Eel photographed on the wreck of the MV *Robert* off Lundy. Image width *c.* 20cm.

Grey Triggerfish *Balistes capriscus* favour places where they can find shelter and use their trigger to jam themselves into a crevice for a doze. Here, on the *James Eagan Layne* in Whitsand Bay, west of Plymouth. Image width *c.* 35cm.

Rosy Featherstars *Antedon bifida* often dominate surfaces of wrecks. Here on ex-HMS *Scylla* in Whitsand Bay, west of Plymouth. Image width *c.* 10cm

Inside the wreck of the SS *Hispania* in the Sound of Mull, off Oban. The still conditions attract active suspension feeders such as the sea squirt *Ascidia mentula* and Yellow-ringed Sea Squirt *Ciona intestinalis* and many surfaces are covered by the sessile stage of Moon Jellyfish *Aurelia aurita*. Image width *c.* 40cm.

Wrecks and other structurally complex habitats provide shelter for many fish species that seem to especially favour them. Here, Bib *Trisopterus luscus* are seen on the wreck of the SS *Rosehill* off south-east Cornwall in July 2014. Image width *c.* 1.5m.

The species that colonise wrecks – but, more particularly, those that do not – may inform estimates of the likely recovery of reef species after, for instance, damage by mobile fishing gear. Although it always has to be borne in mind that some species may not favour steel as a substratum to settle on, there are several species that do not colonise wrecks either because the wrecks are too far away from natural surfaces for the larvae to reach, or because the species reproduces infrequently and grows very slowly.

One of the aft boilers (the fireboxes can be seen at the base) on the wreck of the *Iona II* off Lundy, photographed in June 2014. The vessel sank in 1864. Natural reef habitats, about 1km away, are characterised by large, branching axinellid sponges (*Axinella dissimilis* and *Homaxinella subdola*) and cushion sponges (species of *Polymastia*), together with Pink Sea Fans *Eunicella verrucosa*. None of those sponges have been found to have settled on the wreck, and sea fans are extremely sparse. Image width *c.* 2.5m.

Concrete block and boulder reefs

There have been a very few experimental studies placing concrete blocks on the seabed to answer questions about habitat design (mainly concerning providing suitable habitats for crabs and lobsters). In Poole Bay in 1989, Antony Jensen, Ken Collins and others built a reef of concrete blocks which, after 30 months, had accumulated about 250 different species. Off Lismore in Loch Linnhe near Oban, Tom Wilding and colleagues created several separate reefs of concrete blocks which were designed to study the interaction between man-made structures and their environment, including potential benefits for fisheries and local biodiversity. The communities that developed, attached to and living on the reefs, were mainly alga-dominated and appeared, after a few years, very similar to those found in shallow rocky areas.

Quarried boulders are most often used when building breakwaters and they are de facto artificial reefs, but without any intention of increasing biodiversity or providing habitats for shellfish – even if they do. Breakwaters are shallow structures and most likely to be colonised by algae unless the surrounding waters are highly turbid or the boulders heavily grazed or shaded. In Weymouth Bay, quarried Portland stone has been deposited from a barge in an area licensed for placement of a variety of structures, mainly as an attraction for divers but also with the aim of providing habitat for shellfish.

Above: part of one of the Lismore reefs created near Oban. The reefs were constructed between 2001 and 2006. This image was taken in August 2005. Image width *c.* 1.5m.

The outer (southern) slope of Plymouth Sound breakwater in October 2007. Image width *c.* 1.5m.

The *Scylla* story – not a wreck

Ex-HMS *Scylla* was placed on the seabed in Whitsand Bay, south-east Cornwall, at the end of March 2004. Obtaining and preparing a decommissioned frigate for sinking as an artificial reef was no small job and was led by the National Marine Aquarium (NMA) in Plymouth, with preparation undertaken in the dockyard at Devonport including expert support from the Canadian Artificial Reef Consulting team. *Scylla* has been described as a 'climbing frame for divers' but has been much more than that, especially for the scientific understanding of colonisation processes. Monitoring the colonisation on *Scylla* very nearly didn't happen, but was 'rescued' by the observations and photographs of local divers, especially through the Seasearch programme. In 2006, *Scylla* week, organised by the NMA, included sampling to supplement the observation of conspicuous species. In 2010, with others, I published a description of colonisation of the reef in the *Journal of the Marine Biological Association*. The species count at the end of March 2009 was 263 and only a few additions have been made since then.

Colonisation of the reef showed wide fluctuations in species abundance during the first two years and some surprises. For instance, there were unexpected large settlements of the Green Sea Urchin *Psammechinus miliaris* and of Queen Scallops *Aequipecten opercularis* (both very rarely seen in shallow subtidal areas off Cornwall).

The bridge of ex-HMS *Scylla* in September 2009, four-and-a-half years after the vessel had been placed on the seabed. The reef is visually dominated by Plumose Anemones *Metridium dianthus* with Dead Men's Fingers *Alcyonium digitatum*. Image width *c.* 2.5m.

There were more predictable settlements of tube worms, barnacles and Common Starfish *Asterias rubens*. The impact of grazing by the urchins and starfish was to bring the structure back to a clean battleship grey by spring 2005. However, by 2006, most species that later would dominate or characterise the reef in a 'mature steel wreck community' had settled and the numbers of urchins had been greatly reduced by the arrival of wrasse.

Scylla has taught us an enormous amount about colonisation processes and which species (those that settled readily and grew quickly) were the ones that would recover rapidly from disturbance. Other species that followed-on after the first five years were the ones that would take longer to re-colonise and there are many species that occur on nearby reefs but have not (yet) settled on *Scylla* – the ones that have a potentially poor recovery potential. The story is not a simple one and account has to be taken of whether there are nearby sources of larvae and spores that would reach *Scylla*. Artificial reefs provide opportunities to understand potential for recovery, and monitoring needs to be well-planned and undertaken thoroughly.

The bow of ex-HMS *Scylla* in August 2011.

Left: 30 January 2005 and it's 'not looking good' for a rich wreck community to develop on *Scylla* 10 months after placement. Image width *c.* 2m.

Right: early settlement of tube worms, barnacles and much else had been grazed away by Green Sea Urchins *Psammechinus miliaris*. Also in the image, keeled tube worms *Spirobranchus triqueter* and the Yellow-ringed Sea Squirt *Ciona intestinalis*. 18 February 2005. Image width *c.* 15cm.

Left: but, then, the wrasse and possibly other fish arrived and middens of urchin debris appeared. Settled species were now much less at risk. 15 July 2005. Image width *c.* 18cm.

Right: Common Starfish *Asterias rubens* had quickly consumed early settlements of Blue Mussels *Mytilus edulis*, and would continue to attack later settlements. 14 June 2011. Image width *c.* 50cm.

Below left: a small variety of algal species settled in the first two years on ex-HMS *Scylla* and then a wider variety, including perennials, provided a similar range of species over following years. The only kelp species to establish successfully was Furbelows *Saccorhiza polyschides* (young plants are shown here on 7 January 2017) and other such as Cuvie *Laminaria hyperbor*ea were recorded for the first time with certainty in 2017. Image width *c.* 50cm.

Below right: Orange Pumice bryozoan *Cellepora pumicosa* were first observed in summer 2006 and, unexpectedly, became extremely abundant in the next year. Also in the image, sea squirts *Ascidiella aspersa*: early colonists that persisted for about five years. 19 September 2009. Image width *c.* 50cm.

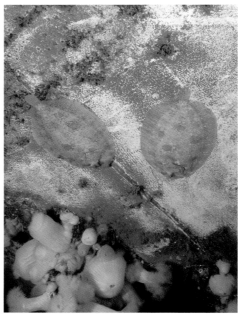

Left: four years and four months after *Scylla* was placed on the seabed, Pink Sea Fans *Eunicella verrucosa* settled. The nearest natural reef populations are about 40m away from the vessel and it seems that reproduction may be sporadic. Photographed on 10 December 2008 when some individuals had grown to 15cm in height. Image width *c.* 40cm.

Right: Ex-HMS *Scylla* was protected from fouling by the now banned Tributyl Tin (TBT) antifouling paint. Attached marine life grows on rust blisters and grills whilst almost nothing else will settle on the painted area. Topknots *Zeugopterus punctatus* nevertheless seem unconcerned and are frequently seen. 26 February 2009. Image width *c.* 70cm.

Spectacular growths of anemones (Plumose Anemones *Metridium dianthus* and pink Elegant Anemones *Sagartia elegans* var. *rosea* with a grey area of the encrusting sea squirt *Diplosoma spongiforme* in this image) colonised the slightly overhanging starboard side of *Scylla*. Here in June 2007, just over three years after the vessel had been placed on the seabed. Image width *c.* 40cm.

Oil or gas rigs and wind farms

There is little published information on marine life colonising offshore structures but there was a significant review in 2003, by Paul Whomersley and Gordon Picken, of the long-term dynamics of fouling communities on installations in the North Sea. A recently launched industry programme (Influence of man-made Structures In the Ecosystem – INSITE) is being undertaken and should provide a much better understanding of colonisation processes and the marine life that occurs on and around artificial structures in the North Sea. I have not tried to arrange dives on offshore structures, on the assumption that any proposal to use recreational scuba diving would give managers and diving supervisors an apoplectic fit and generate a big 'no'. In 2002, with Harvey Tyler-Walters and Hugh Jones, I undertook research for the UK Department of Trade and Industry to advise, amongst much else, upon what sorts of marine life would colonise offshore wind farm structures. Once surveys of newly developed wind farms were undertaken, those predictions proved correct. One such installation is the Walney Offshore Wind Farm and the images used here illustrate two major features of colonisation on steel structures offshore: domination by Blue Mussels *Mytilus edulis* in shallow water, and by Plumose Anemones *Metridium dianthus*, as well as some other anemones and a layer of bryozoans and hydrozoans, in deeper water. The attached fauna was visited especially by crabs. Anti-scour measures (boulders) at the base of the monopiles were colonised by lobsters and crabs and there were shoals of Poor Cod *Trisopterus minutus*.

Species colonising the monopiles in the Walney Offshore Wind Farm. In shallow depths, Blue Mussels *Mytilus edulis* dominate the structures and, with increasing depth, Plumose Anemones *Metridium dianthus* and other attached species take over colonisation. The photographs are taken by a remotely operated vehicle developed by Aquatech-UK. Images: NIRAS Consulting UK / DONG Energy. Image widths *c.* 60cm.

Under the boardwalk – marina pontoons

Whether or not pontoon floats count as seabed (perhaps as upside-down seabed), they are fascinating and often include a rich variety of species. Many of them, however, are likely to be non-natives brought into harbours attached to the hulls of visiting vessels. Suspended above the seabed, the pontoon floats are generally inaccessible to predators such as crabs and starfish, so that mussels especially thrive on them. The pontoon floats in Newlyn, west of Penzance, have been found by David Fenwick to be very rich in species with, during successive samples taken in 2015, 320 identified, of which 4% were non-native and at least 4% were undescribed species or rare British ones. The communities that colonise pontoon floats are not included in the classification of biotopes.

Exploring upside-down seabed at the Mayflower marina in the Tamar Estuary north of Plymouth. Here, the outward parts of the pontoon floats are in sunlight and visually dominated by algae, mainly *Ulva* sp.

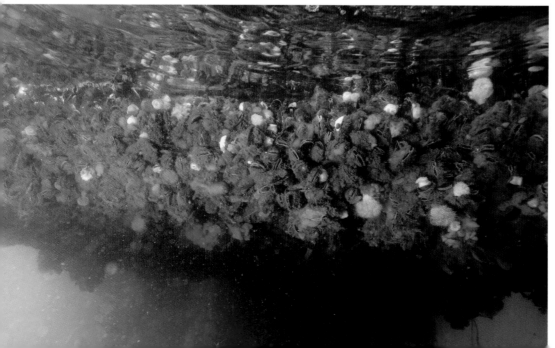

A view from under the boardwalk in a marina. Here, the shaded surfaces are dominated by Blue Mussels *Mytilus edulis* with Plumose Anemones *Metridium dianthus* and encrusting sea squirts. Queen Anne's Battery marina in Plymouth Sound. Image width *c.* 1m.

Some species that are particularly characteristic of marina pontoon habitats:

Left: wherever there are strong currents and full to reduced or variable salinity, tubular hydroids (here, *Ectopleura larynx*) thrive and may dominate the underlying mussels, especially in spring. Image width *c.* 10cm.

Right: algae fringe the sunlit parts of pontoon floats. Non-native Wakame *Undaria pinnatifida* were imported to France as a mariculture species and have spread to many marinas and bays via shipping. Image width *c.* 40cm.

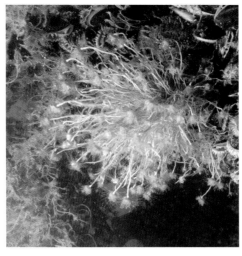

Left: branching bryozoans are often abundant on pontoon floats. This one is a non-native species, *Bugula neritina*. Image width *c.* 5cm.

Right: colonial sea squirts are a feature of pontoon floats. Pictured here, a *Botrylloides* sp. Image width *c.* 5cm.

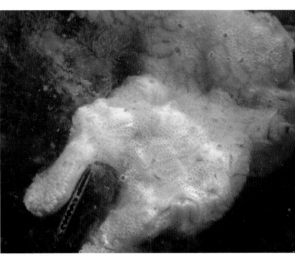

Left: solitary sea squirts thrive in shaded, slightly estuarine waters. Pictured here, the non-native Leathery Sea Squirt *Styela clava*, above, and Yellow-ringed Sea Squirt *Ciona intestinalis* below. Image width *c.* 10cm.

Right: *Perophora japonica*, a non-native sea squirt, has a sporadic appearance on pontoons and elsewhere. The sponge is *Hymeniacidon perlevis*. Image width *c.* 5cm.

Left: a common species on the open coast but also a frequent coloniser on pontoon floats: the Light Bulb Sea Squirt *Clavelina lepadiformis*. Image width *c*. 5cm.

Right: aggregations of Grey Triggerfish *Balistes capriscus*, a summer visitor, may occur in marinas, but not every year. Image width *c*. 60cm.

Left: tubicolous amphipods, *Jassa falcata*, thrive in areas of strong tidal currents. Here, they are on pontoon floats at Tosnos Point in the Kingsbridge Estuary north of Salcombe. Image width *c*. 2cm.

Right: sea slugs (nudibranchs) are attracted by feeding opportunities. Here, Lined Polycera *Polycera quadrilineata* on pontoon floats at Tosnos Point in the Kingsbridge Estuary. Image width *c*. 2cm.

The future of artificial habitats

Oil and gas rigs proliferate in the North Sea and may persist long after the reservoirs they exploit are exhausted. The concept known as 'rigs to reefs' has been greatly pursued in locations such as the Gulf of Mexico and is being promoted in the North Sea as offshore structures related to the oil industry become redundant. Such structures, together with offshore wind farms or other energy devices, increase biodiversity in areas which are otherwise composed entirely of sediment. Near to the coast, they may attract recreational diving and angling. Other benefits include the exclusion of mobile fishing gear in the region of such structures, so that the adjacent seabed becomes a de facto marine protected area. Reefs may be deliberately created as a part of biodiversity offsetting, where a proposed development destroys one habitat and the developer offers compensation in terms of the introduction of another habitat that will increase biodiversity in the area. Sculpture parks may attract divers and help support local businesses that rely on divers. Bereavement balls are made of concrete that incorporates the ashes of a dead loved one who can now sleep with the fishes

Solace Stones are moulded from crushed Portland Stone mixed with cement. They are placed in a licenced area in Weymouth Bay and contain the ashes of a deceased loved-one. They are designed to include fissures and cavities that will attract marine life. Image: Dave Fowler, Cocoa Media.

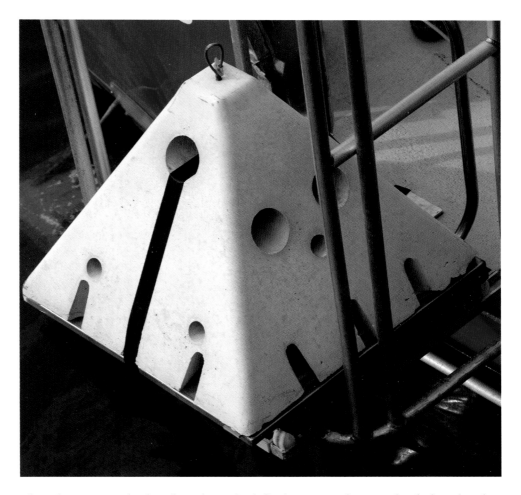

when the structure is placed on the seabed. Such constructions can be designed and placed in a location where they will support as great a variety of species as possible. In Britain, any such placement requires a licence.

Not a benefit but a consideration for ecology: artificial reefs may provide stepping stones that facilitate the spread of species beyond their normal geographical limits, which are determined by barriers such as extensive areas of unsuitable substratum (sediments). Those stepping stones may also assist the spread of invasive non-native species. To an extent, such stepping stones already exist in the form of shipwrecks.

To date, very few structures have been deliberately placed on the seabed around Britain as artificial reefs: perhaps because the coastline is already greatly composed of reefs that support rich biological communities. If there are to be artificial reefs, thought needs to be given to ensuring that any 'placement' is not just a convenient way of disposing of rubbish such as building rubble or car tyres, which are likely to move around or contain toxins that will damage marine life. Well-thought-out artificial reefs could be designed to mimic natural reefs, including the use of stone and the inclusion of structural features such as fissures, overhangs and small caves. Such structures should be used to monitor colonisation and to better understand how communities develop. They should be used to identify which species colonise and, more importantly, for understanding recovery potential, which species do not.

SEA LOCHS

Sea lochs are an iconic part of Scottish scenery. Their surface hides a distinctive and often spectacular underwater landscape of cliffs, shallow tidal rapids and deep mud, with species that thrive only, or mainly, in their wave-sheltered environments. Sea lochs comprise three main types: fjords (long narrow inlets of the sea with a shallow entrance deep inside), fjards (complex systems of water bodies connected by narrows or silled passes) and open sea lochs (fully marine but enclosed areas of sea with no entrance sill or pass). Most are near to fully marine except in their upper reaches, although with variable salinity in the top few metres from freshwater runoff that often forms a dark and peat-stained layer. Loch Etive near Oban receives the greatest freshwater input of any of the large sea lochs in Scotland, giving a marked brackish character to its communities. It is the seabed of fjordic sea lochs that is described here.

Despite their accessibility, whatever the weather, and their distinctive character, the assemblages of species in subtidal areas of sea lochs were little studied until the 1970s and 80s when John Gage, Tom Pearson and colleagues sampled in Loch Etive and Loch Creran north of Oban, and Loch Linnhe and Loch Eil near Fort William. Later, there was a major programme of work undertaken by the University Marine Biological Station at Millport as a part of the Marine Nature Conservation Review and published in 1994. The Millport team (Christine Howson and Rohan Holt especially) brought together results from surveys of 86 sea lochs that included 1,278 sites and revealed, at the time, 60 subtidal biotopes. Descriptive survey work continued and continues to be undertaken, mostly by Scottish Natural Heritage and by Seasearch volunteers.

Sea lochs host distinctive habitats that are dominated by structurally important species. They include those of Horse Mussels *Modiolus modiolus*, maerl (mainly *Phymatolithon calcareum*), Gaping File Shells *Limaria hians*, and serpulid worms *Serpula vermicularis*. They are described later in this chapter, in the 'Species as habitats' section.

The head of Loch Creran: a fjordic sea loch north of Oban. The foreground is 14km from open coastal waters at the Lynn of Lorn.

Top left: lush growths of Sugar Kelp *Saccharina latissima* on shallow bedrock in Loch Carron. Image width *c.* 1.5m.

Top right: in shallow water in many sea lochs, grazing by sea urchins (here, the Edible Sea Urchin *Echinus esculentus*) prevents the establishment of foliose seaweeds. Port à Mheirlich at the entrance to Loch Carron. Image width *c.* 2.5m.

Rocky habitats at or near the entrance to sea lochs

In the parts of sea lochs that are close to open coast in character (full salinity, very or moderately sheltered from wave action and often with strong tidal currents), many of the features that characterise shallow rock habitats described and illustrated in the earlier section, 'Open coast rocky areas', are found. Those habitats include kelp forests, brittlestar beds, rock walls with sponges, sea firs, Dead Men's Fingers, tube worms, crustaceans, encrusting and branching sea mats, starfish, featherstars, sea urchins and sea squirts. There are fish species that live on or near the rocky seabed and that are frequently seen by divers.

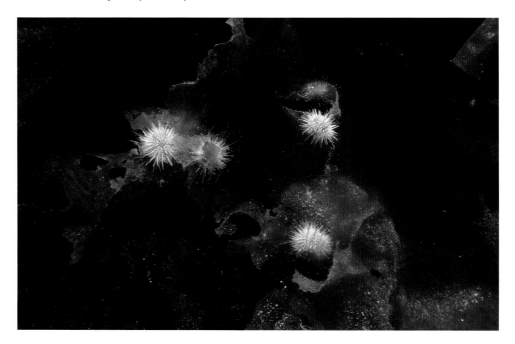

Sea urchins are also able to take advantage of the still conditions in sea lochs and graze on the kelp fronds. Pictured: Green Sea Urchin *Psammechinus miliaris* in Loch Duich. Image width *c.* 25cm..

The shallow infralittoral zone is likely to be dominated by kelps with large fronds that shade the rocks below, reducing the light available for foliose algae to grow. But it is not only shading that affects the opportunity for lush growths of seaweeds to develop – sea urchins thrive in many shallow areas and graze away the algae, preventing development of a rich biota.

In the very tide-sheltered areas of sea lochs, rock surfaces are likely to extend to only shallow depths and may be isolated boulders or cobbles that nevertheless support algal beds. Where tidal currents are, as in this image, moderately strong, for instance on sills and in the narrows at the entrance to sea lochs, kelps are likely to predominate. Cuvie *Laminaria hyperborea* at the entrance to Loch Carron. Image width *c.* 1.5m. Closest to 'Mixed kelp with foliose red seaweeds, sponges and ascidians on sheltered tide-swept infralittoral rock' (A3.222 / IR.MIR. KT.XKT).

Deeper, where animals dominate the rocky habitats, brittlestar beds may develop on the level seabed of cobbles and pebbles, while lush growths of sea firs, sea squirts and featherstars occupy rock slopes.

Left: level seabed of cobbles and pebbles at the entrance to Loch Carron dominated by brittlestars and Dead Men's Fingers *Alcyonium digitatum*. Clumps of Horse Mussels *Modiolus modiolus* are also present in this habitat. Image width *c.* 1m. '*Ophiothrix fragilis* and/or *Ophiocomina nigra* brittlestar beds on sublittoral mixed sediment' (A5.445 / SS.SMx.CMx.OphMx).

Opposite: a rock wall near the entrance to Loch Carron with the yellow Boring Sponge *Cliona celata*, sea firs, the Rosy Featherstar *Antedon bifida* and Edible Sea Urchins *Echinus esculentus*. Image width *c.* 50cm. '*Antedon* spp., solitary ascidians and fine hydroids on sheltered circalittoral rock' (A4.313 / CR.LCR.BrAs.AntAsH).

Some of the fish frequently seen on shallow rock habitats at the entrance to sea lochs. All images taken at the entrance to Loch Carron.

Left: the Leopard-spotted Goby *Thorogobius ephippiatus* is commonly seen under overhangs. Image width *c.* 7cm.

Right: Yarrell's Blenny *Chirolophis ascanii* is most abundant in northern waters and seen mainly in fissures in the rock. Image width *c.* 5cm.

Above: the Fifteen-spined Stickleback *Spinachia spinachia* may be seen swimming amongst the kelp. Image width *c.* 20cm.

Right: the Butterfish *Pholis gunnellus* may be seen swimming through the beds of brittlestars, on kelp stipes or on bedrock. Image width *c.* 10cm.

Shallow sediments and maerl beds

Shallow sediments in parts of sea lochs that are very sheltered from wave action and tidal currents are usually muddy. However, the nature of the seabed is not just due to water movement: it depends also on sediment supply. In some cases there may be very little silt from land runoff and there may be glacial remains of coarse sediment or spillover of hard substrata from the shoreline. Shallow muddy gravel is a widespread habitat in sea lochs as is muddy sand, both supporting a rich community of plants and animals on and within the sediment.

Shallow sediments in sea lochs with little water movement are typically muddy sand with Sugar Kelp *Saccharina latissima* and with burrowing species such as Lugworms *Arenicola marina*. Photographs taken in Loch Linnhe on the approach to Fort William. Image width *c.* 2m and 25cm. '*Saccharina latissima* and *Chorda filum* on sheltered upper infralittoral muddy sediment' (A5.522 / SS.SMp.KSwSS.LsacCho).

In some very sheltered parts of sea lochs, there may be gravelly sediments, such as here near the head of Loch Duich at about 5m depth. The image shows the holdfasts of Sugar Kelp *Saccharina latissima*, Horse Mussels *Modiolus modiolus*, Green Sea Urchins *Psammechinus miliaris*, the tentacles of a terebellid worm and other species. Image width *c.* 35cm. '*Saccharina latissima* with *Psammechinus miliaris* and/or *Modiolus modiolus* on variable salinity infralittoral sediment' (A5.523 / SS.SMp.KSwSS.SlatMxVS).

The communities in Caol Scotnish, a branch of Loch Sween, are unusual. The shallow (less than about 2m below chart datum) sediments are covered by ghostly beds of algae occupied by small Black Brittlestars *Ophiocomina nigra*, with clumps of velvet-like green *Codium* sp. Deeper (but only to about 4m), detached balls of the rarely seen northern maerl species *Lithothamnion glaciale* appear and form beds overrun by Common Brittlestars *Ophiothrix fragilis*. Other unattached maerl species present in Britain occur more widely, including in open coast areas and tidal sounds, but beds of detached *L. glaciale* appear confined to a very few sea lochs.

The shallow, muddy slope in Caol Scotnish (here at about 1m below chart datum) is colonised by Bootlace Weed *Chorda filum*, filamentous algae and Black Brittlestars *Ophiocomina nigra*, which often climb up the Bootlace Weed. In the centre is a clump of red alga topped by Common Brittlestars *Ophiothrix fragilis*. Image width *c.* 1.5m. Closest to: 'Mixed kelp and red seaweeds on infralittoral boulders, cobbles and gravel in tidal rapids' (A3.223 / IR.MIR.KT.XKTX).

Beds of detached Northern Maerl *Lithothamnia glaciale* occur in only a few sea lochs. Here, it is pictured in Caol Scotnish (Loch Sween), with the Common Brittlestar *Ophiothrix fragilis*. Image width *c.* 20cm. '*Lithothamnion glaciale* maerl beds in tide-swept variable salinity infralittoral gravel' (A5.512 / SS.SMp.Mrl.Lgla).

Deeper mixed substrata

Whilst the sides of many sea lochs are mud slopes, there are often rock outcrops, isolated boulders and occasional debris such as lost lobster creels. Those hard substratum surfaces are colonised by a fauna characteristic of sea lochs, although not necessarily the same in different locations or even between nearby boulders. For instance, the species of solitary sea squirts that dominate or characterise the surfaces vary with location: sometimes there are brittlestars and sometimes not; sometimes there are serpulid tube worms forming reefs and sometimes not, and so on.

A mud slope with scattered boulders and bedrock outcrops extends to the mud plain characterised by sea pens in Loch Linnhe on the approach to Fort William. Rocks with Sea Loch Anemones *Protanthea simplex*, brittlestars and sea squirts next to a Phosphorescent Sea Pen *Pennatula phosphorea* in mud. Image width *c.* 35cm. Closest to: '*Novocrania anomala* and *Protanthea simplex* on very wave-sheltered circalittoral rock' (A4.3141 / CR.LCR.BrAs.NovPro.FS).

An oasis of hard substratum life with sponges, sea firs, Sea Loch Anemones *Protanthea simplex*, tube worms, solitary ascidians and other animals on a sediment slope with burrowing anemones in the background, Loch Linnhe. Image width *c.* 30cm. Biotope as above.

On boulders where silt settles on the upward-facing surface, it will be the sides that are colonised by typical sea loch species. Here, at Ardgarton in Loch Long, Sea Loch Anemones *Protanthea simplex*, sea squirts *Ascidia virginea*, Common Brittlestars *Ophiothrix fragilis* and tubes of Peacock Worms *Sabella pavonina* are seen. The brachiopod *Novocrania anomala* is present but scarcely visible. Image width *c.* 40cm. '*Novocrania anomala* and *Protanthea simplex* on sheltered circalittoral rock' (A4.314 / CR.LCR.BrAs.NovPro).

Stony muddy sediments in depths greater than about 10m attract a community found in only a few sea lochs and here most conspicuously characterised by the sea cucumber *Psolus phantapus*. Upper Loch Creran. Image width *c.* 45cm. 'Sparse *Modiolus modiolus*, dense *Cerianthus lloydii* and burrowing holothurians on sheltered circalittoral stones' (A5.442 / SS.SMx.CMx.ClloModHo).

Beds of the colourful sea cucumber *Psolus phantapus* occur in only a few places in moderate depths where there are cobbles or pebbles embedded in the sediment. The body is U-shaped and juveniles attach to hard substrata by a 'sole' where the tube feet are concentrated. They are believed to appear above the sediment only in May and June.

The sea cucumber *Psolus phantapus* in stony muddy sediment near the head of Loch Duich. This colourful photograph is a stark contrast to the text book illustrations of contracted and preserved specimens. Image width *c.* 10cm.

Deep mud

Swim down to the level plains of deep mud in the lochs and the seabed takes on an ethereal appearance. In the dim light, the diver will see forests of sea pens and stunningly beautiful Fireworks Anemones *Pachycerianthus multiplicatus*. The sediment is burrowed by worms and by crustaceans, which may show themselves at the entrances. The most likely burrowing crustacean to be seen is the Rugose Squat Lobster *Munida rugosa*: it is the 'lobster' you will eat if you buy a bowl of lobster tails. If you are lucky, there will be Norway Lobsters *Nephrops norvegicus* at the entrance to their burrows, perhaps accompanied by Fries's Goby *Lesueurigobius friesii*. But the crustaceans that make the most complex burrows will be out of sight. The mud shrimps *Calocaris macandreae* and *Callianassa subterranea* live deep in the mud, often in anoxic conditions. There will be smaller Burrowing Anemones *Cerianthus lloydii*, burrowing brittlestars – most likely *Amphiura filiformis* – and other starfish, such as the northern *Solaster endeca*. A night dive may be particularly exciting as, with torches turned off and a bit of a prod, the Phosphorescent Sea Pens *Pennatula phosphorea* are said to 'sparkle like the lights on a Christmas tree'.

Deep muddy sediment at the head of Loch Duich with the Fireworks Anemone *Pachycerianthus multiplicatus* in the foreground and Tall Sea Pens *Funiculina quadrangularis*. Image width *c.* 1m in the foreground. 'Seapens, including *Funiculina quadrangularis*, and burrowing megafauna in undisturbed circalittoral fine mud' (A5.3611 / SS.SMu.CFiMu.SpnMeg. Fun).

Inhabitants of deep muddy sediments

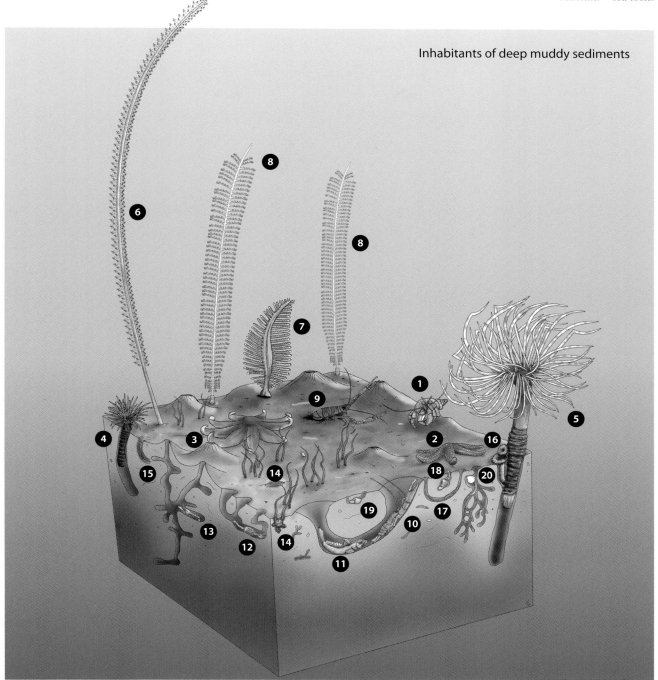

Based on the seabed near the head of Loch Duich. On the surface of the sediment are **1** a hermit crab *Pagurus bernhardus*, **2** the Common Starfish *Asterias rubens* and **3** the Purple Sunstar *Solaster endeca*. Burrowing species that show at the surface are **4** the Burrowing Anemone *Cerianthus lloydii* and **5** the Fireworks Anemone *Pachycerianthus multiplicatus*, **6** the Tall Sea Pen *Funiculina quadrangularis*, **7** the Phosphorescent Sea Pen *Pennatula phosphorea* and **8** Slender Sea Pens *Virgularia mirabilis*. Mobile species that live in burrows are **9** the Rugose Squat Lobster *Munida rugosa* and **10** the Norway Lobster *Nephrops norvegicus*, together with **11** Fries's Goby *Lesueurigobius friesii*. Mostly hidden from view but with connections with the surface are the burrowing mud shrimps **12** *Calocaris macandreae* and **13** *Callianassa subterranea*, and **14** brittlestars *Amphiura filiformis*. There will be other burrowing species, especially polychaete worms such as **15** *Terebellides stroemii*, **16** *Myxicola infundibulum*, **17** *Glycera* sp. and bivalve molluscs such as **18** *Corbula gibba*, **19** *Nucula sulcata* and **20** *Thyasira flexuosa*. 'Seapens, including *Funiculina quadrangularis*, and burrowing megafauna in undisturbed circalittoral fine mud' (A5.3611 / SS.SMu.CFiMu.SpnMeg.Fun). Drawing: Jack Sewell.

The three species of sea pens found in Scottish sea lochs:

Slender Sea Pen *Virgularia mirabilis* in Loch Linnhe. Image width *c.* 25cm.

Phosphorescent Sea Pen *Pennatula phosphorea* in Loch Linnhe. Image width *c.* 20cm.

Tall Sea Pen *Funiculina quadrangularis* in Loch Duich. Image width *c.* 80cm.

The Fireworks Anemone *Pachycerianthus multiplicatus*. Image width *c.* 20cm.

Species that live in burrows. **Left:** Norway Lobster *Nephrops norvegicus* at the mouth of its burrow in Loch Shieldaig north of Loch Carron. Image: Paul Naylor. **Right:** nearby, a Rugose Squat Lobster *Munida rugosa*, that, here, prefers a discarded glass jar to a burrow. Image widths *c.* 15cm.

Rock slopes and walls

It is the rock walls in sea lochs that provide some of the most spectacular dives in Scotland, with remarkable species that create highly characteristic biotopes extending to depths in excess of 50m. Rock walls are often going to be in locations where there are strong currents – and in Loch Duich, just about 1km east of the iconic Eilean Donan castle, where the photographs that follow were taken, the current always seemed, for me, to be running significantly!

Shallow depths are likely to be grazed heavily by sea urchins *Psammechinus miliaris* and *Echinus esculentus* with only sparse kelps *Laminaria hyperborea*. Deeper, rocks become colonised by brittlestars *Ophiopholis aculeata*, *Ophiocomina nigra* and *Ophiothrix fragilis*. Look more closely and the diver will see high densities of brown limpet-like shells about 10mm across. These are brachiopods or lampshells *Novocrania anomala*. Brachiopods have been in our seas for over 550 million years but are now a very minor part of our fauna. Deeper, and not in such high numbers, another brachiopod species, *Terebratulina retusa*, may be found. Scattered amongst the dominant species may be large colonies of Football Sea Squirts *Diazona violacea* and other ascidians. The high density of Peacock Worms *Sabella pavonina* creates a spectacular sight, but only if they are extended, which often they are not. Amongst the worm tubes will be large numbers of the Sea Loch Anemone *Protanthea simplex*. Those who have explored deeper than about 35m, and down to 55m depth, will have found the Celtic Featherstar *Leptometra celtica*, together with the more widely distributed Rosy Featherstar *Antedon bifida*.

Shallow depths with plants of Cuvie *Laminaria hyperborea* over heavily grazed rock. Image width *c.* 2m. 'Grazed *Laminaria hyperborea* park with coralline crusts on lower infralittoral rock' (A3.2144 / IR.MIR. KR.Lhyp.GzPk).

Below: a step in the steeply sloping rock with Sea Loch Anemones *Protanthea simplex*, a Common Whelk *Buccinum undatum*, Light Bulb Sea Squirts *Clavelina lepadiformis*, a Gas Mantle Sea Squirt *Corella parallelogramma* and the Yellow-ringed Sea Squirt *Ciona intestinalis*. The rocks are largely covered in dark red encrusting algae and pink encrusting algae. Image width *c.* 30cm.

A variant of the *Novocrania anomala* and *Protanthea simplex* community on vertical rock at about 20 to 25m depth where the cliffs are festooned with Peacock Worms *Sabella pavonina*, with rocks otherwise covered in Sea Loch Anemones *Protanthea simplex*, serpulid worms and, scarcely visible, brachiopods *Novocrania anomala*. Image width *c.* 1m. '*Novocrania anomala* and *Protanthea simplex* on very wave-sheltered circalittoral rock' (A4.3141 / CR.LCR.BrAs.NovPro.FS).

Below: a variant of the *Novocrania anomala* and *Protanthea simplex* community on overhanging rock at about 30m depth. Here, Sea Loch Anemones *Protanthea simplex* and the Yellow-ringed Sea Squirt *Ciona intestinalis* are especially abundant. The northern Red Cushion Star *Porania pulvillus* is also present. Image width *c.* 70cm.

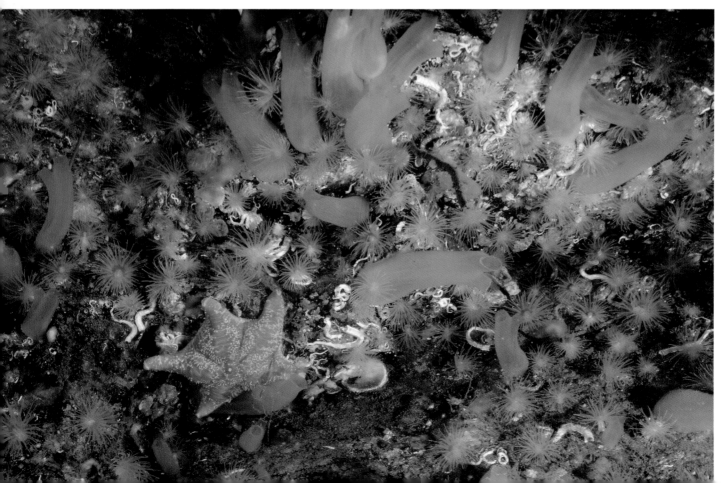

Above: a variant of the *Novocrania anomala* and *Protanthea simplex* community on vertical rock at about 40m depth with the brachiopods clearly visible, together with brittlestars *Ophiocomina nigra* and *Ophiothrix fragilis* and a Football Sea Squirt *Diazona violacea*. Image width *c.* 70cm.

Right: two species of brachiopod found on rock surfaces in sea lochs. The common one is *Novocrania anomala* (left) and, less often seen, *Terebratulina retusa* (right). Image widths *c.* 2cm.

RIAS AND VOES

Rias are long, narrow inlets of the sea that were once river valleys, and are a feature of south-west Britain. The sides are often steep and rocky. Voes and firths (in Shetland and Orkney) are similarly long, narrow inlets that most likely started as small river valleys that were further deepened by glaciers and eventually flooded after the last glaciation. Almost none of the voes have the entrance sills that characterise many sea lochs: Ronas Voe is an exception and has two sills and rocky sides, making it more like a fjordic sea loch. The volume of freshwater that enters a ria or voe varies from one to another but is usually very low in proportion to the seawater component. Where high, significantly reduced salinity is usually confined to the upper reaches or to surface water. The marine life that develops reflects the stability brought about by lack of extreme wave action, although tidal currents may be strong in narrow sections.

Weisdale Voe on the west coast of Shetland, photographed in 1986. The voe is over 40m deep below chart datum.

The River Fal at King Harry Ferry north of Falmouth, photographed in April 2017. The flooded river valley at this location is up to 16m deep below chart datum and the bottom water close to full salinity. King Harry Ferry is 8km from the open sea at Pendennis Point, Falmouth.

The rias of south-west Britain were extensively described as part of the 'Harbours, Rias and Estuaries' study commissioned by the Nature Conservancy Council (NCC) in the 1980s and carried out by the Field Studies Council Oil Pollution Research Unit (OPRU). Voes in Shetland were a part of the research commissioned by the Institute for Terrestrial Ecology and carried out by Bob Earll and Chris Lumb in the mid-1970s. Later, in the mid-1980s, Shetland was surveyed by OPRU to contribute to the NCC's Marine Nature Conservation Review of Great Britain. Marine inlets, whether estuaries or rias and voes, are attractive as ports and some of the most extensive and detailed investigations of seabed sediment communities have been undertaken as part of monitoring studies around the oil terminals in Milford Haven, commissioned especially by various oil companies and, in Sullom Voe, through the Shetland Oil Terminal Environmental Advisory Group (SOTEAG).

Rocky habitats

It is the rocky habitats that particularly distinguish rias and, to an extent, voes. The species that thrive in conditions of full or variable salinity and extreme shelter from wave action (but often strong tidal currents) give the reefs a characteristic appearance. Some widely distributed open coast species such as the Boring Sponge *Cliona celata*, antenna hydroids *Nemertesia* spp., Dead Men's Fingers *Alcyonium digitatum* and Jewel Anemones *Corynactis viridis* persist, but some such as branching Yellow Staghorn Sponge *Axinella dissimilis*, cushion sponges *Polymastia* spp., Pink Sea Fans *Eunicella verrucosa*, Yellow Cluster Anemones *Parazoanthus axinellae*, the Edible Sea Urchin *Echinus esculentus*, Cotton Spinner *Holothuria forskali*, Ross 'coral' *Pentapora foliacea* and sea squirts *Stolonica socialis* are likely to be seen rarely or not at all in south-west rias. Echinoderms (spiny-skinned animals) are thought to not favour areas of variable salinity but the rocky seabed in some rias can be covered in Rosy Featherstars *Antedon bifida* and, especially where there are beds of mussels, Common Starfish *Asterias rubens* are likely to be present.

Varied seaweed-dominated communities persist in rias, although often turbid conditions mean that they may not extend deeper than about 5m below chart datum level. Here, at Scoble Point in Salcombe Harbour, Golden Kelp *Laminaria ochroleuca* occurs around about chart datum level. Image width *c.* 60cm.

Above: anastomosing colonies of the Shredded Carrot Sponge *Amphilectus fucorum* may dominate areas of seabed in ria habitats. Image width *c.* 6cm.

Above right: the sponge *Halichondria bowerbanki* is typical of ria and some other wave-sheltered habitats. Here housing an Edible Crab *Cancer pagurus*. Image width *c.* 20cm.

Left: the branching sponge *Raspailia ramosa* is characteristic of ria communities. Image width *c.* 15cm.

Right: Mermaid's Glove *Haliclona oculata* is a characteristic sponge of wave-sheltered habitats exposed to moderate to strong currents, including in rias. Behind the sponge is the antenna hydroid *Nemertesia antennina* and various other sponges. Image width *c.* 20cm.

Left: in spring, athecate (lacking a sheath) hydroids such as *Garveia nutans* and, in the middle, *Tubularia indivisa*, may dominate parts of the tide-swept areas of rias. Image width *c.* 4cm.

Right: Jewel Anemones *Corynactis viridis* may thrive in the tide-swept parts of rias. Here they are pictured with the Double Spiral Worm *Bispira volutacornis*. Image width *c.* 6cm.

Above: clusters of the Coral Worm *Salmacina dysteri* are a frequent feature of rias. Image width *c.* 3cm.

Right: Rosy Featherstars *Antedon bifida* may visually dominate rocky areas in rias. Image width *c.* 50cm.

155

Shallow rock communities at Pembroke Ferry east of the Cleddau Bridge in Milford Haven where salinity on the seabed is variable but sometimes drops below 30. The species characteristically include Breadcrumb Sponge *Halichondria panicea*, *H. bowerbanki*, Shredded Carrot Sponge *Amphilectus fucorum*, Mermaid's Glove sponge *Haliclona oculata*, the branching sponge *Raspailia hispida*, Goosebump Sponge *Dysidea fragilis*, Coral Worm *Salmacina dysteri* and the Finger Bryozoan (or Sea Chervil) *Alcyonidium diaphanum*. Image width *c.* 80cm. Identified as 'Cushion sponges, hydroids and ascidians on turbid tide-swept sheltered circalittoral rock' (A4.2511 / CR.MCR.CFaVS.CuSpH. As), but may be 'Cushion sponges and hydroids on turbid tide-swept variable salinity sheltered circalittoral rock' (A4.2512 / CR.MCR. CFaVS.CuSpH.VS).

In a similar situation to the image from Pembroke Ferry, a slightly different (perhaps more stable and close to full salinity) community, but identified as the same biotope, at Eastern King Point in Plymouth Sound. Here, Dead Men's Fingers *Alcyonium digitatum*, a sponge *Suberites* sp., colonial sea squirts *Distomus variolosus* and white Sandalled Anemones *Actinothoe sphyrodeta* are most conspicuous. Image width *c.* 40cm. 'Cushion sponges, hydroids and ascidians on turbid tide-swept sheltered circalittoral rock' (A4.2511 / CR.MCR. CFaVS.CuSpH.As).

The submerged river gorge that forms part of the Plymouth Sound and estuaries ria obstructs tidal flow and creates strong currents over its lip. Here, a typical ria rock community is shown, with Poor Cod *Trisopterus minutus* hanging over the top of the escarpment to catch passing plankton. Image width *c.* 40cm.

Sediment habitats

Rias and voes often have extensive areas of muddy or sandy sediments and there may be a rich diversity of species present in and on those sediments. For instance, in a review commissioned by SOTEAG of the very many surveys of sediment macrobenthos (from grab samples sieved over a 1mm mesh) in and around Sullom Voe between 1977 and 2014, around 1,270 species had been identified. The communities that occur in those sediments are not unique to rias and voes and may also occur in open coast bays sheltered from strong wave action such as in Torbay, Carrick Roads at the entrance to the Fal and Mounts Bay in west Cornwall. However, on the deep level seabed in some of the central channels, strong currents may lead to the presence of coarse muddy sediments not often seen elsewhere. Maerl and Horse Mussel, *Modiolus modiolus*, beds may be a feature of tide-swept parts of rias and voes and are described in the section on *Species as habitats*.

Very shallow sediments in the open parts of rias and voes are often characterised by seagrass *Zostera marina* – but not necessarily the lush growths of more open coast locations such as those found in the Isles of Scilly. Here, in late January at Firestone Bay, Plymouth Sound, isolated clumps of seagrass are seen with foliose red algae attached to their rhizomes at about 1m below chart datum. In the centre are plants of Wireweed *Sargassum muticum*. Image width *c.* 60cm. '*Zostera (Zostera) marina* beds on lower shore or infralittoral clean or muddy sand' (A5.5331 / SS.SMp.SSgr.Zmar).

Level or gradually sloping sandy sediments with shell material may have conspicuous species present on the surface, such as these Auger Shells *Turritella communis*, but the richest variety of species is below the surface. Photograph taken at West Hoe in Plymouth Sound. Image width *c.* 40cm. '*Melinna palmata* with *Magelona* spp. and *Thyasira* spp. in infralittoral sandy mud' (A5.334 / SS.SMu.ISaMu. MelMagThy).

Inhabitants of muddy sediments within rias and voes

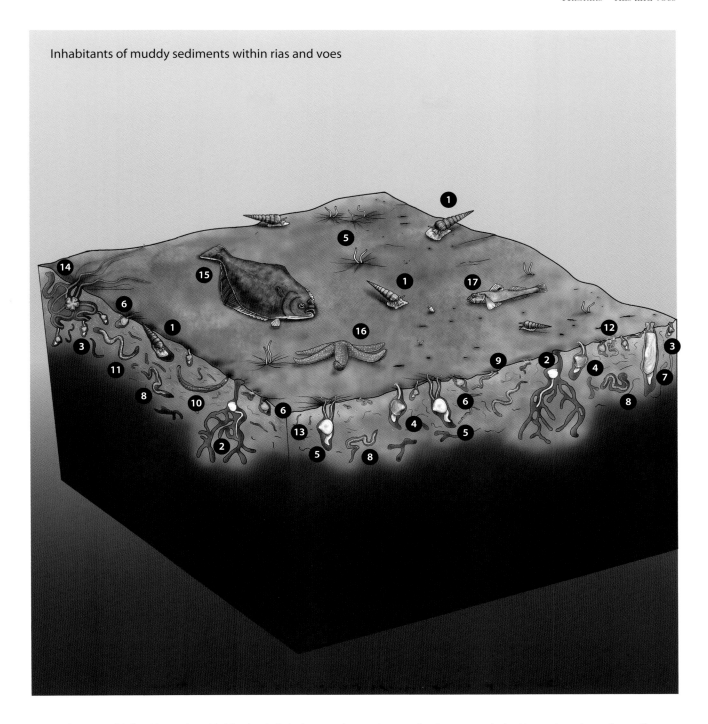

The southern part of Sullom Voe at about 20–30m depth. Only the upper few centimetres of sediment are colonised by most species as, deeper, the sediment becomes deoxygenated. The main species illustrated are **1** Turret Shells *Turritella communis*, the bivalve molluscs **2** *Thyasira flexuosa*, **3** *Kurtiella bidentata*, **4** *Corbula gibba*, **5** *Abra alba*, **6** *Nucula nitidosa* and **7** *Phaxas pellucidus*, along with worms **8** *Hilbigneris gracilis*, **9** *Prionospio fallax*, **10** *Glycera alba*, **11** *Scoloplos armiger* and **12** *Chaetozone setosa*. There will be many other small species, especially worms and **13** amphipods and polychaetes, also present in a rich community with as many as 130 species recorded from three 0.1m² grab samples. On the surface are **14** brittlestars *Amphiura chiajei*, **15** the Common Dab *Limanda limanda*, **16** the Common Starfish *Asterias rubens* and **17** a sand goby *Pomatoschistus* sp. Based on various sources. 'Kurtiella bidentata and *Thyasira* spp. in circalittoral muddy mixed sediment' (A5.443 / SS.SMx.CMx.KurThyMx). Drawing: Jack Sewell.

Part of a deep muddy shell gravel community sampled by dredge and including the starfish Sand Star *Astropecten irregularis* and Common Starfish *Asterias rubens*, the bivalve mollusc *Phaxas pellucidus*, tubes of the polychaete worm *Lagis koreni*, the Harbour Crab *Liocarcinus depurator* and the Slender Sea Pen *Virgularia mirabilis*. Location: Vaila Sound, Shetland. Image width *c.* 25cm. Image: Keith Hiscock/JNCC. Most likely '*Lagis koreni* and *Phaxas pellucidus* in circalittoral sandy mud' (A5.355 / SS.SMu.CSaMu.LkorPpel).

Some species photographed in ria and voe habitats but also found in shallow wave-sheltered sediments in harbours or bays:

A bubble snail *Scaphander lignarius* in the Helford River. The shell is normally covered in mucus and is seen ploughing through the surface sediment. On the right, the mucus cover has been removed. Its English name is the Woody Canoe-bubble. Image width *c.* 8cm.

Left: the hydroid *Corymorpha nutans* is a seasonal coloniser of shallow sediments but is prey to some sea slugs. Here, it is the sea slug *Fjordia browni* in early May in Plymouth Sound at the entrance to the River Tamar. Image width *c.* 5cm.

Right: only the parchment tubes of the worm species *Chaetopterus* show above the sediment, but may visually dominate shelly substrata. Here they are at Millbay Pit in Plymouth Sound. Image width *c.* 4cm.

Left: the Fountain Anemone *Sagartiogeton laceratus* on a muddy slope in Voe of Clousta, Shetland. Image width *c.* 12cm.

Mixed sediments in voes are likely to have sparse Horse Mussels *Modiolus modiolus* . Here, they are seen together with brittlestars, an Edible Sea Urchin *Echinus esculentus* and the sediment starfish Sand Star *Astropecten irregularis*. Location: Busta Voe, Shetland. Image width *c.* 40cm.

In the centre of ria channels, the level seabed is usually tide-swept and composed of muddy shell and rock gravel colonised by a small variety of species. Here, most conspicuously, two Neptune's Heart sea squirts *Phallusia mammillata* and a hermit crab *Pagurus bernhardus*, photographed at a depth of 11m below chart datum off Scoble Point in Salcombe Harbour. The empty bivalve shells are most likely the Pullet Carpet Shell *Venerupis corrugata*. Image width *c.* 40cm. 'Circalittoral mixed sediments' (A5.44 / SS.SMx.CMx).

Rarely, isolated parts of voes become deoxygenated near to the seabed during summer. This deoxygenation may occur only in occasional years and is documented for the inner basin of Sullom Voe by Tom Pearson and Tasso Eleftheriou. The isolation of the water is brought about by the establishment of a thermocline. In Ronas Voe, which has the characteristics of a fjordic sea loch, the presence of sills further exacerbates isolation of the basins.

Deoxygenated conditions in one of the basins of Ronas Voe, shown here in late August 1986, destroy vulnerable species (the white shells are probably of the bivalve mollusc *Corbula gibba*) and lead to the presence of bacterial mats. Image width *c.* 50cm. '*Beggiatoa* spp. on anoxic sublittoral mud' (A5.7211 / SS.SMu.IFiMu.Beg).

A tide-swept rock wall in the Falls of Lora near Oban, western Scotland, where tidal currents can commonly exceed 6 knots. Here, rocks are colonised by barnacles *Balanus crenatus* which have been consumed by Dog Whelks *Nucella lapillus* (also in the image), leaving just their calcareous bases. There are tubular hydroids *Tubularia indivisa*, Dahlia Anemones *Urticina felina*, Breadcrumb Sponges *Halichondria panicea* and Hornwrack *Flustra foliacea* most obviously present. Image width *c.* 50cm. '*Balanus crenatus* and *Tubularia indivisa* on extremely tide-swept circalittoral rock' (A4.111 / CR.HCR.FaT.BalTub).

TIDAL SOUNDS AND OTHER EXTREMELY TIDE-SWEPT HABITATS

Tide-swept habitats can occur off exposed headlands and where two land masses come close together. In tidal channels, the seabed is more likely to be sheltered from wave action but subject to strong tidal flow as tidal currents are squeezed between the land. Where the strength of tidal currents is less than about 3 knots (about 260cm/sec) at the surface, a great many species thrive. Between 3 and 6 knots, a smaller number of species will thrive and the assemblages that develop become typical of tide-swept rocks. In a very few locations around Britain, tidal currents exceed 6 knots and the assemblages will consist of a few highly tolerant species including mobile species that could be described as 'keeping their heads down' while the strongest current is running.

Tide-swept channels are also a feature of rias and of sea lochs, which have their own sections in this chapter. However, in general, this section describes extreme conditions where currents in excess of 6 knots at the surface affect seabed habitats.

Above opposite: between the bridges crossing the Menai Strait, tidal currents can reach an impressive 8 knots and seabed biotopes characteristic of tide-swept conditions are found. To the left (east) is the Telford suspension bridge, towards the right (west) the island of Ynys Gorad Goch and, to the far right, the Britannia rail and road bridge.

The field work that I undertook for my PhD targeted locations subject to different strengths of wave action and tidal currents. Most of my work in tide-swept channels was in the Menai Strait. As a result of those field studies, I could see points along the spectrum of tidal current velocity where changes occurred in the seabed assemblages.

Shallow tide-swept habitats at Ynys Gorad Goch in the Menai Strait. Kelp extends only to shallow depths, mainly because of the high turbidity of the water and because suspension-feeding animals out-complete foliose algae for space. The wide angle picture (image width c. 1m) emphasises encrusting sponges Shredded Carrot Sponge *Amphilectus fucorum* and Breadcrumb Sponge *Halichondria panicea*, and barnacles *Balanus crenatus*. The close-up image (image width c. 25cm) begins to reveal other characteristic species: tubular hydroids *Tubularia indivisa* and Elegant Anemones *Sagartia elegans*, as well as an overall covering of muddy tubes and a fuzz of Linear Skeleton Shrimps *Caprella linearis*. 'Laminaria hyperborea' park with hydroids, bryozoans and sponges on tide-swept lower infralittoral rock' (A3.2122 / IR.MIR.KR.LhypT.Pk).

The muddy tubes that often make underwater rocks look more like a farmyard in tide-swept habitats are of jassid amphipods (*Jassa falcata* here) and the fuzz on the rocks is of caprellid amphipods, also known as skeleton shrimps (Linear Skeleton Shrimp *Caprella linearis* here). Both species are likely to be out and about while the current is slight to moderate, crawling around and facing their feeding appendages into the water flow to catch food. As the strength of the current increases, they hunker down (the caprellids) or retreat into their tubes (the jassids) to avoid being swept away. Photographed at the Britannia rail and road bridge in the Menai Strait. Image widths *c.* 2cm.

The tide-swept habitats of the Menai Strait were the location for one of the first systematic surveys of seabed assemblages. In 1956, Wyn Knight-Jones, W. Clifford Jones and Doreen Lucas made observations along a transect just east of the Telford suspension bridge. Twenty years after this original survey, Wyn Knight-Jones and Tony Nelson-Smith took samples at different depths along the transect. They identified 231 species including 27 algae. The transect was again systematically sampled in 1976 by Richard Hoare and Martina Peattie. I dived the site frequently during my PhD studies in the early 1970s and most recently in June 2015. The dominant and characterising species were remarkably similar in the descriptions that spanned 58 years and are broadly as described in the next paragraph.

The Telford suspension bridge over the tidal current-exposed Menai Strait. A transect extending from the shore in the foreground to the opposite side was first surveyed in 1956, then again in 1976 and recent observations were made by the author in 2015. The seabed assemblages described in the two detailed surveys were as observed in 2015 and examples are shown below.

Sponges (the branching sponge *Haliclona oculata* and Shredded Carrot Sponge *Amphilectus fucorum*) with a turf of branching sea mats and sea firs. Image width *c.* 50cm. '*Flustra foliacea* and *Haliclona (Haliclona) oculata* with a rich faunal turf on tide-swept circalittoral mixed substrata' (A4.137 / CR.HCR.XFa.FluHocu).

Sea firs (mainly *Abietinaria abietina*, *Sertularia argentea* and *Hydrallmania falcata*) with Finger Bryozoan (or Sea Chervil) *Alcyonidium diaphanum* and a variety of other species. Image width *c.* 50cm. 'Very tide-swept faunal communities on circalittoral rock' (A4.11 / CR.HCR.FaT).

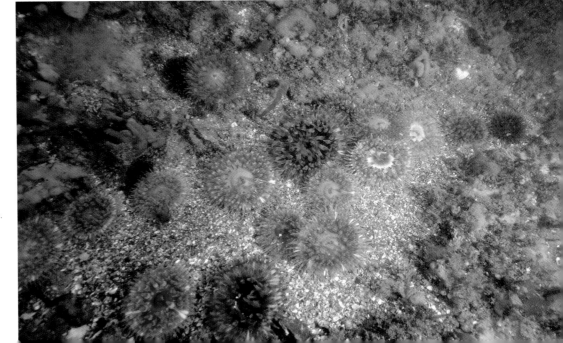

Rock covered in a thin layer of shell gravel with Dahlia Anemones *Urticina felina*. Image width *c.* 50cm. '*Urticina felina* and sand-tolerant fauna on sand-scoured or covered circalittoral rock' (A4.213 / CR.MCR. EcCr.UrtScr).

The Falls of Lora under Connell Bridge at the entrance to Loch Etive photographed on a neap tide! Current velocity exceeds 6 knots on spring tides with reports of 10 knots by kayakers with a GPS. The Scottish Association for Marine Science had a flow meter on the bottom that recorded speeds of 12 knots. Only a small number of species can survive on the rocky seabed. A popular site for a drift dive by experienced scuba divers.

The Breadcrumb Sponge *Halichondria panicea* and tubular hydroids *Tubularia indivisa* growing together are especially characteristic of tide-swept habitats. Other species include often high abundances of Elegant Anemones *Sagartia elegans* and, in or near to clean gravel, Dahlia Anemones *Urticina felina*. The hydroids *Sertularia argentea* or *Sertularia cupressina* (the two are very similar in appearance) and *Hydrallmania falcata* are often visually dominant. Encrusting over the rocks, barnacles *Balanus crenatus* may cover large areas, but there will be patches where only the bases remain – the barnacles are prey to Dog Whelks, *Nucella lapillus*, usually considered an intertidal

Animals that benefit from the plentiful food supply brought by strong tidal currents in the narrows at the entrance to sea lochs may also out-compete algae, even in the absence of sea urchins where, in the location illustrated, low or variable salinity make it an unsuitable habitat for urchins. Here, in the Falls of Lora, rocks are dominated by barnacles *Balanus crenatus*, Breadcrumb Sponge *Halichondria panicea* and tubular hydroids *Tubularia indivisa*. Image width *c.* 60cm. '*Balanus crenatus* and *Tubularia indivisa* on extremely tide-swept circalittoral rock' (A4.111 / CR.HCR.FaT.BalTub).

species but often abundant wherever there are barnacles to be consumed. Hornwrack *Flustra foliacea* is another species that might form dense patches and Blue Mussels *Mytilus edulis* sometimes form beds. In the absence of strong wave action and with a plentiful food supply, some sponges that are normally thin encrustations on the open coast may form thick cushions with tall (chimney like) or anastomosing growth forms. They include Breadcrumb Sponge *Halichondria panicea*, Shredded Carrot Sponge *Amphilectus fucorum* and the often purple sponge *Haliclona cinerea*.

Domination of rocks by the barnacle *Balanus crenatus* attracts a high abundance of Dog Whelks *Nucella lapillus* that feed on them. The Shore Crab *Carcinus maenas* (on the rock shelf) is typical of areas with low and variable levels of salinity. Here in the Falls of Lora Image width *c.* 60cm.

Plumose Anemones *Metridium dianthus* have very wide limits of tolerance in relation to strength of wave action and tidal currents and often thrive in high energy environments. Here, they are pictured in the Falls of Lora. Image width *c.* 40cm.

This sunken rescue dummy was discovered in a gulley at the Falls of Lora and already has fast-growing hydroids *Sertularia argentea*, and barnacles *Balanus crenatus*, typical of high energy environments, growing on it. The gravel is of Dog Whelk *Nucella lapillus* shells.

The most extreme of tidal sounds must be the whirlpool of Corryvrekan at the north end of the island of Jura where, in addition to the currents being squeezed between two land masses, there is an underwater pinnacle that rises from 130m and accelerates the tidal flow even more. Surface tidal currents are described as up to 10 knots. I have dived the Corryvreckan (on the north side of Jura) once. My main recollection is of watching shallow kelp pushed flat onto the seabed in one direction and, about half a minute later, flat on the seabed in the opposite direction. I achieved a bottom time of nine minutes before the current became too strong. When I looked up survey records for the location, it was remarkable that my dive, with Annette Little and Sarah Fowler in June 1982, was the only description of the narrows – many other marine biologists have dived the sound known as 'the Corry' and there is an excellent video by David Ainsley of the site. Two biotopes were recorded: '*Laminaria hyperborea* park and foliose red seaweeds on tide-swept lower infralittoral mixed substrata' (A3.2132 / IR.MIR. KR.LhypTX.Pk) and '*Balanus crenatus* and *Tubularia indivisa* on extremely tide-swept circalittoral rock' (A4.111 / CR.HCR.FaT.BalTub).

The sites with 'extreme' tidal current that I have mentioned in this section should only be dived with thorough planning and by experienced divers.

Horseshoe Rocks off Bull Point in north Devon is in an area that is exposed to tidal currents of up to 3 knots but the obstruction created by the rock pinnacle accelerates those currents, creating overfalls. Although the maximum current strength is unknown, the assemblages that develop on the rock are characteristic of open coast extremely tide-exposed habitats. Here, in the main picture, there are patches of Blue Mussels *Mytilus edulis*, Elegant Anemones *Sagartia elegans*, Dahlia Anemones *Urticina felina*, tubular hydroids *Tubularia indivisa* and a Common Starfish *Asterias rubens*. Other characteristic species include Plumose Anemones *Metridium dianthus* and Jewel Anemones *Corynactis viridis* (left image), Hornwrack *Flustra foliacea* (middle image) and, in the right-hand picture, the tubular hydroids extended by what are described as 'whips' believed to be of the amphipod crustacean *Dyopedos porrectus* with, on the seabed, the colonial tunicate *Synoicum incrustatum*. Image widths *c.* 80cm for the main image and *c.* 25cm for closer images. Includes: '*Tubularia indivisa* on tide-swept circalittoral rock' (A4.112 / CR.HCR.FaT.CTub); '*Mytilus edulis* beds with hydroids and ascidians on tide-swept exposed to moderately wave-exposed circalittoral rock' (A4.241/ CR.MCR.CMus.CMyt).

ESTUARIES

Estuaries have been used and abused by humans for millennia and most host towns, cities and ports. They were the main transport highways before roads, leading to the presence of ports and quays, many of which are now disused. Estuaries are important for commerce, industry, fisheries, wildlife, recreation, defence and for disposal of waste. Many have been intensively sampled and described for their biology. Due to ever-changing levels of salinity, estuaries tend to support a low diversity of species, but with their nutrient-rich inputs are typically highly productive and support high populations of those species that can survive in them. Only a few species are truly estuarine; most are brackish-water-tolerant marine or freshwater organisms. Few are attractive to divers and underwater photographers!

Estuaries are areas where freshwater from rivers meets and mixes with seawater. This zone of mixing moves up and down the estuary with the tide and with other factors, especially river flow. They are divided into three main zones according approximately to critical salinity tolerance limits for marine species: fully marine (30–35); reduced (18–30) and low (<18) [the numerical values are approximately equivalent to parts per thousand of salts]. Identifying the salinity characteristics at any location in an estuarine system must be done after sampling in conditions ranging from drought to prolonged heavy rain. Also, since freshwater is less dense than seawater, the salinity at the seabed in estuaries is likely to be higher than at the surface through formation of a phenomenon known as a salt-water wedge, where incoming seawater pushes along the seabed underneath the freshwater, allowing brackish-water-intolerant species to reach further and migrate up into the estuary with the tides. Most readings of salinity will have been taken at the surface – marine species may penetrate much further up the estuary on the seabed than is thought possible according to those surface readings.

Estuaries are often highly turbid habitats and the subtidal sampling in them is usually done remotely by grab, dredge or trawl. In most estuaries, the seabed comprises muddy sediments with shells, clinker and other debris. In some, strong currents may expose the underlying bedrock or clay. The sediments present at a location vary greatly according to river and tidal flows, sediment inputs and other factors. Often, sediments are covered by terrestrial debris in the form of leaves and branches. The consequent high proportion of organic detritus supports high populations of bacteria and a rich meiofauna, but with mud frequently anoxic just 1cm below the surface. Bedrock reefs occur in the Severn Estuary and in ria estuaries such as the Daucleddau Estuary at the head of Milford Haven, in the River Fal north of Falmouth, in the Kingsbridge Estuary north of Salcombe Harbour and the Tamar Estuary inland of Plymouth Sound.

Fjordic sea lochs can rarely be described as estuaries, but Loch Etive north of Oban has a very high freshwater input and has many estuarine characteristics in its shallow waters (say less than 15m depth). Brackish-water-tolerant species such as Blue Mussels *Mytilus edulis*, the Shore Crab *Carcinus maenas* and the Baked Bean Ascidian *Dendrodoa grossularia* colonise those shallow depths in the loch. Others species that tolerate low and variable salinity are those illustrated in the section on 'Tidal sounds and other extremely tide-swept habitats' where the reefs at the entrance to Loch Etive are illustrated. Echinoderms are notably sensitive to low salinity and, in Loch Etive, the Green Sea Urchin *Psammechinus miliaris* and the Edible Sea Urchin *Echinus esculentus* do not occur.

Inhabitants of estuarine mud

Based on parts of the Tamar Estuary. The species living in the mud are the worms **1** *Nephtys hombergii*, **2** *Aphelochaeta marioni*, **3** *Capitella capitata* and **4** *Tubificoides* sp. along with the bivalve molluscs **5** *Cerastoderma edule* and **6** *Abra nitida*. Mobile species on the mud are **7** the Shore Crab *Carcinus maenas*, **8** the Common Goby *Pomatoschistus microps*, **9** the Flounder *Platichthys flesus*, **10** the Brown Shrimp *Crangon crangon* and **11** a shrimp *Palaemon longirostris*. 'Aphelochaeta marioni and Tubificoides spp. in variable salinity infralittoral mud' (A5.322 / SS.SMu.SMuVS.AphTubi). Additionally, a patch of terrestrial debris has been colonised by **12** Blue Mussels *Mytilus edulis*, and by **13** an estuarine hydroid *Hartlaubella gelatinosa* with an associated swarm of **14** mysid shrimps. Swimming over the bottom is **15** the Sea Bass *Dicentrarchus labrax* and foraging in the mud, **16** the Thick-lipped Grey Mullet *Chelon labrosus*. Drawing: Jack Sewell.

Sediments in estuaries are affected by the very variable conditions there. The rippled nature of the sandy mud here at Cargreen, 11km upstream from the entrance to the Tamar Estuary at Plymouth, suggests mobile sediments most likely colonised by ephemeral species. The shells of a Common Cockle *Cerastoderma edule*, are present. Image width *c.* 25cm.

Sediments

Most studies of subtidal sediment habitats in estuaries have been in relation to pollution, especially in the vicinity of sewage or industrial effluents. A significant feature of estuarine sediments is that they have a tendency to accumulate contaminants from the overlying water and act as long-term reservoirs of pollutants long after the original inputs have ceased. Contextual studies, such as those carried out in the Forth Estuary west of Edinburgh, in the Humber Estuary, in some of the Essex estuaries, and in the River Dee between north Wales and England, record a very limited sediment fauna characterised by polychaete worms such as *Streblospio shrubsolii*, *Nepthys* spp., *Scoloplos armiger*, *Tharyx* spp., *Ampharete grubei*, *Melinna palmata* and *Eteone longa*, amphipods including *Gammarus* spp. and *Corophium* spp. There may

Estuarine mud in an area of variable salinity at Saltash, 7km upstream from the entrance to the Tamar Estuary at Plymouth and at about 5m below chart datum. Image width *c.* 25cm. The communities here were identified from samples as '*Nephtys hombergii* and *Tubificoides* spp. in variable salinity infralittoral soft mud' (A5.323 / SS.SMu.SMuVS.NhomTubi) or '*Aphelochaeta marioni* and *Tubificoides* spp. in variable salinity infralittoral mud' (A5.322 / SS.SMu.SMuVS.AphTubi).

be bivalve molluscs, particularly where salinity remains relatively high: *Abra* spp. and *Limecola balthica*. Closer to the brackish–freshwater interface, seabed sediments may become dominated by small worms *Polydora* spp. and oligochaete worms. Mobile species such as Brown Shrimp *Crangon crangon*, mysid shrimps and Shore Crabs *Carcinus maenas*, are often a feature of estuarine seabeds.

Estuaries and rias are often important habitats for rays – here a Thornback Ray *Raja clavata* in the Tamar Estuary. Image width *c.* 80cm.

Due to international shipping, estuaries may be hotspots for non-native species – ships and yachts from all over the world berth and bring with them species from similar port environments on their hulls and in ballast water. However, in the case of the Slipper Limpet *Crepidula fornicata*, mariculture was almost certainly the introductory agent. It was in 1894 that the species was first noted in the River Crouch in Britain, following the introduction of American Oysters *Crassostrea virginica*. Slipper Limpets have since become extremely widespread and often visually dominant on subtidal sediment seabeds, from estuaries to some areas of open coast.

The sedimentary bed of some estuaries has become dominated by the non-native Slipper Limpet *Crepidula fornicata*. Here, at the entrance to St John's Lake, 2km upstream from the entrance to the Tamar Estuary at Plymouth, where salinity on the seabed at 15m depth below chart datum only occasionally drops below 30. Image width *c.* 15cm. Most likely the biotope 'Crepidula fornicata and Mediomastus fragilis in variable salinity infralittoral mixed sediment' (A5.422 / SS.SMx.SMxVS.CreMed).

Reef habitats

Hard substrata in estuaries have often been introduced by human activity, although natural rock outcrops may occur in some. Sunken trees and other terrestrial debris may also be important as solid substrata. Sedentary and sessile species on hard substrata in environments with variable or low salinity are a mixture of brackish-water-tolerant species that also occur in the open sea, and a small number of specialists that are either especially abundant in estuaries or that only occur in conditions of low salinity. For this section, I have selected photographs from the Tamar Estuary where outcropping rock can be found throughout its length, but mainly in a few locations where tidal currents are strong. There are other estuaries that have reef habitats and that have been surveyed by diving – especially the Daucleddau in west Wales.

The Baked Bean Ascidian *Dendrodoa grossularia* often dominates hard substratum in estuaries. Here, it is pictured at the entrance to the Tamar Estuary. Image width *c.* 60cm. 'Circalittoral faunal communities in variable salinity' (A4.25 / CR.MCR.CFaVS) – there is no finer level classification for biotopes dominated by *D. grossularia*.

Exploring reef habitats in estuaries is not an attractive prospect for most divers, but can be achieved by choosing times when there has been little or no rainfall for several weeks, during high water when coastal water may predominate, and at the end of neap tides when sediment suspended by strong currents should have settled. The variety of species is highest in the outer parts of an estuary, where salinity may be variable or even low (but not lower than 18) on occasions and many open coast species can survive. Here, there may be sufficient light for shallow waters to be characterised by kelps and by a small range of foliose algae. In deeper areas, attached species will be mostly encrusting sponges with hydroids and ascidians, often with groves of the Peacock Worm *Sabella pavonina*. Sheet-forming colonial ascidians such as species of the Family Didemnidae may also grow over other species. Echinoderms, which are generally sensitive to low salinity, may still occur with the Common Starfish *Asterias rubens* feasting on beds of mussels. Other species of echinoderms such as Rosy Featherstars *Antedon bifida*, Common Brittlestars *Ophiothrix fragilis* and Spiny Starfish *Marthasterias glacialis* may also be present. In tide-swept habitats, there may be clusters of the Plumose Anemones *Metridium dianthus*. Bizarrely, two species that often characterise reefs in low or variable salinity in the shelter of estuaries, also characterise extremely wave-exposed habitats on the open coast: Plumose Anemones *Metridium dianthus* and Baked Bean Ascidians *Dendrodoa grossularia*. The biotopes that occur in variable salinity are often outliers of those described in the section on rias in areas where salinity is close to that in the sea. Moving to the higher reaches of an estuary where salinity will range widely between almost freshwater at times, but up to 30 at high water when there has been no rainfall, the most likely dominant species to be found are Blue Mussels *Mytilus edulis* and specialists such as the hydroid *Hartlaubella gelatinosa* and the encrusting bryozoan *Conopeum reticulum*. Native and Pacific Oysters (*Ostrea edulis* and *Magallana gigas*) also occur in low and variable salinity. In 1986, as a part of the Harbours, Rias and Estuaries Survey of south-west Britain, I dived as far into the Tamar Estuary as underwater visibility allowed – about 15km from the entrance at Plymouth – to find slabs of rock at 2–3m below chart datum dominated by the estuarine/fresh water hydroid *Cordylophora caspia* with *Balanus* sp. and the encrusting bryozoan *Einhornia crustulenta*.

A 'grove' of Peacock Worms *Sabella pavonina* with a Dahlia Anemone *Urticina felina* attached to an upturned dinghy in the Tamar Estuary. Image width *c*. 25cm.

Reef communities at Saltash in the Tamar Estuary 7km upstream from the entrance to the Tamar at Plymouth. Here, salinity at the surface can drop below 10 but, on the seabed, is likely to fall to 24 in the deepest part of the estuary at around 18m below chart datum. Much of the hard substratum is colonised by the biotope 'Cushion sponges and hydroids on turbid tide-swept variable salinity sheltered circalittoral rock' (A4.2512 / CR.MCR. CFaVS.CuSpH.VS).

A rocky outcrop at about 11m below chart datum characterised by Plumose Anemones *Metridium dianthus*, Sandalled Anemones *Actinothoe sphyrodeta*, encrusting sponges *Halichondria* sp. and *Hymeniacidon perlevis*, and the branching sponge Mermaid's Glove *Haliclona oculata*. Image width *c.* 80cm.

Overhang community at about 15m below chart datum characterised by solitary sea squirts Yellow-ringed Sea Squirt *Ciona intestinalis* and *Styela clava* and encrusting didemnid sea squirts. Image width *c.* 30cm.

Mussel bed with Common Starfish *Asterias rubens* and tubular hydroids, characteristic of tide-swept habitats, here at about 10m below chart datum. Image width *c.* 25cm in the foreground.

Typically in estuaries, hard substratum is often of rubbish – in this case a discarded traffic cone colonised by long sponges *Sycon ciliatum*, a few Rosy Featherstars *Antedon bifida*, tubes of Peacock Worms *Sabella pavonina*, keeled tube worms *Spirobranchus triqueter* and a sea fir that is most likely *Sertularia cupressina*.

Reef habitats in estuaries are rarely extensive and these steeply sloping and vertical surfaces in the River Tamar, whilst surrounded by natural rock, are foundations for the rail bridge. The surfaces, here at about 8m below chart datum, are colonised by a variety of cushion and branching sponges and by extensive mats of didemnid sea squirts (most likely *Trididemnum* sp.). A Goldsinny Wrasse *Ctenolabrus rupestris* is foraging amongst the attached species. Shallower, on the bridge supports, Blue Mussels *Mytilus edulis* entirely cover vertical surfaces and are, in turn, colonised by patches of sea firs, encrusting sponges and encrusting sea squirts.

Hard substratum in estuaries may be of terrestrial debris such as sunken trees. Here, at Saltash in the Tamar Estuary, the branches of a tree are colonised by a Dahlia Anemone *Urticina felina*, sea firs most likely *Sertularia cupressina*, a Common Whelk *Buccinum undatum* laying eggs, and a small sea slug. Photographed in early January.

Here, at Cargreen in the Tamar Estuary, 11km upstream from the entrance to the Tamar at Plymouth and at a depth of about 2m below chart datum, salinity is likely to drop below 10 and not rise above 30. Near to the shore, there are Blue Mussel beds *Mytilus edulis* encrusted with barnacles *Balanus crenatus,* with the Shore Crab *Carcinus maenas*, together with the characteristic estuarine hydroid *Hartlaubella gelatinosa*. Image widths *c.* 10cm. '*Hartlaubella gelatinosa* and *Conopeum reticulum* on low salinity infralittoral mixed substrata' (A3.363 / IR.LIR.IFaVS.HarCon). (*C. reticulum*, a characteristic estuarine bryozoan, was found at Cargreen attached to a discarded bottle.)

The non-native hydroid *Cordylophora caspia* may be abundant on hard substrata near the limit of seawater influence on salinity. The example here was collected from Cothele Quay 21km upstream from the entrance to the Tamar Estuary at Plymouth. Image width *c.* 2cm. One of three species characterising the biotope 'Cordylophora caspia and *Einhornia crustulenta* on reduced salinity infralittoral rock' (A3.362 / IR.LIR. IFaVS.CcasEin).

The importance of estuaries

The importance of estuaries for biodiversity goes well beyond their use as feeding areas for birds. There are specialised seabed organisms that occur only or especially in estuaries, which are mentioned earlier in the text. Amongst fish species, few are truly estuarine (spending all their life in the estuary and breeding there), but one is the bottom-living Common Goby *Pomatoschistus microps*. Some species, such as the Flounder *Platichthys flesus*, live most of their lives in estuaries in southern Britain but migrate to the open sea to breed. Many marine fish species use estuaries as sheltered, nutrient-rich nursery areas, often concentrating within narrow zones such as the highly productive freshwater–saltwater interface. These fish include Herring *Clupea harengus*, Sea Bass *Dicentrarchus labrax* and Common Sole *Solea solea*.

Left: Common Sole *Solea solea* enter estuaries to breed. This one was photographed at Saltash in the Tamar Estuary. Image width *c.* 20cm.

Right: Flounder *Platichthys flesus* live most of their time in estuaries but migrate to the open sea to breed. Location: Saltash in the Tamar Estuary. Image width *c.* 20cm.

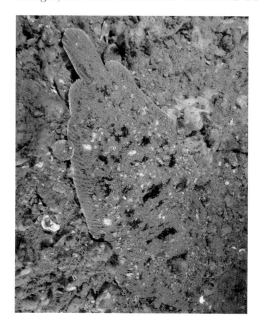

SALINE LAGOONS

Saline lagoons are not the most exciting of habitats to survey, but their shallow seabeds include many unique features. Many lagoon surveys were commissioned by the NCC in the 1970s and 1980s, and the MNCR had a team that surveyed lagoons in Scotland during the 1990s. It was during those MNCR surveys of isolated lagoons that I discovered it is not just the army that goes 'yomping'. Access to some of the lagoons either needed a helicopter (beyond our budget) or involved packing wetsuit, snorkelling equipment and camera gear into a rucksack, putting a weightbelt around our waists and setting off on foot across the boggy ground – sometimes for several miles. I was fortunate in that my participation in these surveys was brief, but yompers like Roger Covey, Frank Fortune, Tim Hill and Kate Northen tackled the challenges enthusiastically and often.

Fundamentally, a lagoon is an area of shallow salty water separated from the sea – so, lagoons can include fully saline habitats (for instance, the 'Blue Lagoon' – Abereiddy Quarry). In a lagoon, except near any entrance, tidal currents are likely to be negligible.

The most recent definition of a saline lagoon allows for brackish, fully saline or hypersaline water bodies to be included although, at one time, a saline lagoon was considered to be a lagoon with reduced salinity, usually less than 30. Saline lagoons are mostly very shallow (less than 5m deep). They may be completely isolated by a barrier of rock or sediment or partially separated by a barrier of shingle, pebbles or small boulders through which percolation occurs. They may have a direct connection with the sea via a channel and sill or a culvert or pipe. Roger Covey observes, 'There are few stranger experiences than coming across what is apparently a peat-fringed upland pool, only to realise that its gravel shore has patches of fucoid seaweeds, and is in fact an arm of the sea extending inland'.

In 1991, Roger Bamber and others estimated that there are 40 known specialist lagoon species in Britain. By 2015, that number stood at 39 in an article by Sue Chambers and others in the *Porcupine Marine Natural History Society Bulletin*. Although referred to as specialist lagoon species, some also occur in estuarine waters or very deep waters and some are non-native species.

Species that occur in saline lagoons are typically those that can survive variable and/or low salinity. They include many species that occur in very sheltered but fully saline locations, such as voes and lochs, and in estuaries. Rocky sides of saline lagoons will most likely support fucoid and green algae. The sediment floors are colonised by seagrass beds of *Zostera noltei* and *Zostera marina* together with, typically in lagoons, *Ruppia maritima* and *R. spiralis*, and often the nationally scarce foxtail stoneworts *Lamprothamnium papulosum* and *Chara aspera*. Fennel Pondweed *Stuckenia pectinata* (previously *Potamogeton pectinatus*) occurs in lagoons where sea water is greatly diluted, typically where salinity is lower than 4. Animals that are typical of saline lagoons include some crustacea and molluscs. In lagoons of higher salinity, seabed communities will be similar to those of extremely sheltered areas on the open coast, with Three-spined Sticklebacks *Gasterosteus aculeatus* likely to be found swimming amongst the seagrass and, on the seabed, Lugworm *Arenicola marina*, sea slugs *Akera bullata*, *Aplysia punctata* and *Philine aperta*, and burrowing amphipods *Corophium volutator*. There are often ascidians (sea squirts) which, because they create their own feeding current, can thrive in such still habitats.

A flooded landscape on North Uist in the Outer Hebrides. Here, a view towards Loch Obasaraigh with Eval to the right. Photographed in 1990.

Scotland has the largest number of lagoons in Britain, especially in the flooded landscapes of the Outer Hebrides. They appear on maps under a variety of titles, for instance 'obs' (rock-bound lagoons in the Hebrides), 'lochs' and 'lochans', and 'houbs' (in Shetland), etc. Studies of Scottish lagoons have been summarised by Stewart Angus who, in 2016, identified 103 saline lagoon habitats covering 33.29km² out of a total of 47.50km² for the whole of Great Britain.

Some of the best developed lagoons are the brackish habitats of the Outer Hebrides, where there is the highest concentration of these habitats. They have attracted the attention of biologists and, in the 1930s, Edith Nicol and M. D. Dunn described the flora and fauna of several lochs. After a 40-year gap in studies, D. H. N. Spence and others surveyed Loch Obasaraigh (Obisary), an unusually deep brackish sea loch connected to Loch Euphort. There followed a series of surveys in the 1970s, commissioned by the NCC and undertaken, in the subtidal zone, by Frances Dipper, Roger Mitchell, Bob Earll and others. Loch Obasaraigh on North Uist is an exceptional location. It is a true silled lagoon, is 3.4km long, and has a maximum depth of 46m, although the greater part of the loch is less than 9m deep. Shallow depths with low salinity (13–14) were characterised by algae and, surprisingly, by a wide range of invertebrates including sea urchins and starfish that are usually sensitive to low salinity, along with dense populations of the Slender Sea Pen *Virgularia mirabilis* with the Burrowing Anemone *Cerianthus lloydii*. There were some areas of Fennel Pondweed *Stuckenia pectinata*. Deeper than the halocline at about 4m, where salinity was just below 30, there were many ascidians and brittlestars with large populations of the opisthobranch sea slug *Akera bullata*.

The Vadills in Shetland is an outstanding example of a complex lagoon system and includes 61 hectares of channels and basins with a rich complement of biotopes supporting a large number of species – and even a loch (Marlee Loch), most likely named after the Shetland word for seagrass (Marlie). In amongst filamentous green algae, beds of the sea cucumber *Leptopentacta elongata* with phoronid worms and the anemone *Edwardsia claparedii* were present. In Orkney, the Loch of Stenness is the outstanding lagoon habitat. At 860 hectares, it is the largest lagoon in Britain, with a very wide range of biotopes that include extensive areas of Fennel Pondweed *Stuckenia pectinata*, seagrass *Ruppia maritima* and Alternate Water-milfoil *Myriophyllum alterniflorum*, together with the lagoon specialist snail *Potamopyrgus antipodarum* (previously *P. jenkensi*, now known to be the non-native New Zealand Mud Snail).

In the Outer Hebrides, 72 lagoons were described as a part of the MNCR, ranging in size from Loch Portain at 0.06ha to Loch Bee at 850ha in extent – one of the largest lagoons in Britain. However, many were of close to full salinity. The MNCR reported 52 marine biotopes or sub-biotopes from the Outer Hebrides lagoons.

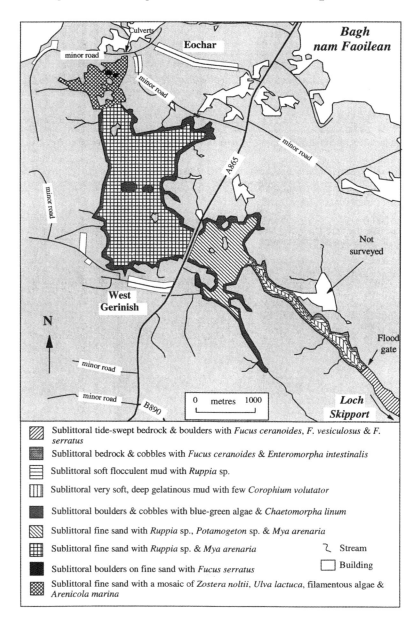

Loch Bee is the largest saline habitat in the Outer Hebrides, with a maximum length of 8km and an area of 850ha. Salinity is between about 17 at northern end to near full salinity (35) at the southern at the time surveys were undertaken. Assemblages typical of saline lagoons occur in the loch. The map is from the MNCR report on saline lagoons in Scotland. Reproduced with permission.

In England, the total area of saline lagoons is some 1,300ha with 450ha being occupied by a single site (The Fleet in Dorset). The photograph above shows a view from Langton Herring east towards Portland, with Chesil Bank seawards.

English saline lagoons are concentrated on the south-east coast and predominantly occur behind a sediment barrier. There, 164 lagoon habitats with a total area of 13.38km² were recorded. East Anglia, in particular, is noted for the presence of a large number of saline lagoons holding rare and scarce lagoon specialist organisms.

Surveying a bed of *Zostera marina* with small amounts of *Ruppia maritima* in The Fleet as part of the Community Seagrass Initiative in September 2016.

Nationally scarce foxtail stonewort, *Lamprothamnium papulosum*, amongst seagrass *Zostera marina* in The Fleet. Images widths *c.* 35cm and 10cm. 'Angiosperm communities in reduced salinity' (A5.54 / SS.SMp.Ang).

The Fleet is a shallow tidal inlet 13km long and connected to the sea through a channel (with strong tidal currents) at the eastern end. There is also seawater percolation through Chesil Bank so that most of The Fleet is polyhaline (salinity of between 18 and 30) to full salinity. The Fleet has been described by Richard Barnes as 'the finest example of a lagoon of its type within the British Isles' and is the only English lagoon that I have studied. The area has been extensively surveyed over many years, and those studies continue as condition assessments to report on the conservation status of the Site of Special Scientific Interest and Special Area of Conservation. All of the Fleet is very shallow and extensively colonised by seagrass. The nationally rare Defolin's Lagoon Snail *Caecum armoricum* occurs in The Fleet, together with other species typical of lagoon habitats such as the Starlet Sea Anemone *Nematostella vectensis*, the crustaceans *Gammarus insensibilis* and *Idotea chelipes*, the sea slug *Akera bullata*, the lagoon cockle *Cerastoderma glaucum*, and the snails *Ecrobia ventrosa* and *Rissoa membranacea*.

Welsh saline lagoons encompass 15 locations with a total area of 0.833km². Cemlyn lagoon on Anglesey is considered the best example of a saline lagoon in Wales. The lagoon is separated from the sea by a shingle bank with a narrow channel and sluice system through which most seawater exchange occurs. There are several specialist lagoon species that occur at Cemlyn including the bryozoan (sea mat) *Conopeum seurati*, the lagoon cockle *Cerastoderma glaucum*, the lagoonal mud-snail *Ecrobia ventrosa* and the lagoonal isopod *Idotea chelipes*. Plants include the Brackish Water-crowfoot *Ranunculus baudotii* and Beaked Tasselweed *Ruppia maritima*.

SPECIES AS HABITATS

Many species provide homes for other species, usually in an accidental way (they are just another surface to attach to or to crawl over). Some provide a structurally complex habitat, offering nooks and crannies for species to find hiding places or shelter from adverse conditions as well, often, as a stable substratum in otherwise mobile sediments. Some are hosts to species which are only, or almost only, found on or within them. Six species that form structurally complex habitats, and that host similar communities wherever they occur, are described on the following pages. Further examples, where a species seems to require a particular host and both derive benefit, and others, where there is a clear advantage for one species but no obvious advantage for the host, are also included.

Seagrass beds

Seagrass beds are a dominant feature of some shallow sandy seabed habitats in wave-sheltered conditions. Before the 1930s, they were much more widely distributed throughout the north-east Atlantic than now. In the 1920s and 1930s, disease and nutrification seemed to be the cause of loss from, in many cases, whole bays and inlets. Those causes have most likely been much reduced, but there has been no significant recovery.

A bed of seagrass *Zostera marina* at St Martin's in the Isles of Scilly. A stalked jellyfish can be seen top left and a bleached clump of Harpoon Weed *Asparagopsis armata* in the centre. Image width *c.* 40cm. '*Zostera* (*Zostera*) *marina* beds on lower shore or infralittoral clean or muddy sand' (A5.5331 / SS.SMp. SSgr.Zmar).

The most prevalent species on subtidal seabeds in full or slightly variable salinity is *Zostera marina*. The plants are angiosperms: they have roots (rhizomes), produce seeds and form continuous beds. The beds are important in stabilising sediments, enabling their colonisation by a greater variety of species than occur on nearby bare sediments. They also provide shelter for young fish, i.e. they have a nursery function. Some species live specifically on the leaves, such as the rare hydroid *Laomedea angulata*. Others occur on leaves of seagrass as well as on foliose algae, for instance stalked jellyfish; or on hard substratum, for instance Snakelocks Anemones *Anemonia viridis*.

Seagrass beds are a specialised habitat that occurs within 'Shallow sandbanks slightly covered by seawater all of the time' and in 'Lagoons', listed as in need of protection in the European Commission's Habitats Directive.

Left: Snakelocks Anemones *Anemonia viridis* are widely distributed on hard substrata. In sedimentary habitats, sea grass leaves provide a surface for attachment. Carricknath, Falmouth Harbour. Image width *c*. 8cm.

Right: stalked jellyfish are often associated with the leaves of seagrass. Here, *Calvadosia campanulata* at St Martin's, Isles of Scilly. Image width *c*. 3cm.

Left: Red Speckled Anemone *Anthopleura ballii* attaches to the rhizomes of seagrass. It is pictured here at St Martin's, Isles of Scilly. Image width *c*. 7cm.

Right: Sea Hares *Aplysia punctata* are often found in seagrass meadows. This one is at Chickerell in The Fleet, Dorset. Image width *c*. 7cm.

In the kelp forest off the southwest coast of Lundy. This picture was taken in 1972 and remains my photograph that gives the best impression of the 'atmosphere' of a kelp forest. Image width *c.* 1m. '*Laminaria hyperborea* with dense foliose red seaweeds on exposed infralittoral rock. (A3.115 / IR.HIR. KFaR.LhypR).

Kelp forests

Kelp forests occur around the world in cold temperate habitats. 'Kelp' originally referred to the ashes of seaweed used as a source of potash and iodine but, in British waters, the term is commonly used to describe large, brown algae that form a canopy over underlying rock or consolidated sediment. The subtidal plants that particularly attract associated assemblages of species are Cuvie *Laminaria hyperborea*, Golden Kelp *L. ochroleuca* and Furbelows *Saccorhiza polyschides*.

Many scuba divers avoid kelp and may be easily spooked if they find themselves trying to push through the forests. But, to a marine biologist, kelp forests are a fascinating habitat that attracts a wide range of species, some specific to the fronds, stipes or holdfasts.

Kelp fronds are especially attractive to the sea fir *Obelia geniculata* and the encrusting sea mat *Membranipora membranacea*. The rough stipes of Cuvie are favoured by the red algae *Palmaria palmata*, *Phycodrys rubens* and *Membranoptera alata* and, especially in the absence of grazing by sea urchins, attract a rich seaweed assemblage. However, Golden Kelp has smooth, clean stipes and appears to be very effective at repelling species that may attach to them. Holdfasts are not just a point of attachment for the plants, but can provide shelter from predators for mobile species such as amphipod crustaceans and worms as well as an attachment point for sessile species. As mentioned on page 34, in 'Gathering knowledge and creating information', over 300 species have been recorded attached to and living in the holdfasts of Cuvie, although the fauna associated with Golden Kelp holdfasts is much less diverse.

Kelp forests occur on hard substrata and on Reefs, a habitat listed in the EC Habitats Directive.

By the end of the summer, Cuvie *Laminaria hyperborea* plants are heavily colonised with sea firs (hydroids) and sea mats (bryozoans) on the fronds and a variety of algae on the rough stipes. In contrast, the fronds and stipes of Golden Kelp *Laminaria ochroleuca* are much cleaner-looking. The images are from rock outcrops south of the Plymouth Sound breakwater and in the Landing Bay at Lundy: both photographed in mid-August. Both species are perennial. Image widths *c.* 50cm. 'Laminaria hyperborea with dense foliose red seaweeds on exposed infralittoral rock' (A3.115 / IR.HIR.KFaR.LhypR) and 'Mixed *Laminaria hyperborea* and *Laminaria ochroleuca* forest on moderately exposed or sheltered infralittoral rock' (A3.1153 / IR.LIR.K.LhypLoch).

Furbelows *Saccorhiza polyschides* at the end of the summer and in a sheltered (silty) location. The bulbous holdfast leads to the convoluted reproductive structures (the 'furbelows') and onto the broad frond. This is a large plant that will soon begin to break up and be swept away or become food for grazing snails as winter approaches. Image width *c.* 1.2m.

Horse Mussel reefs

Horse Mussels *Modiolus modiolus* are widely distributed around Britain, but only reach a significant size and only form biological (biogenic) reefs northwards from north Wales to parts of the west coast of Scotland and especially in Orkney and Shetland. Horse Mussel beds are (and have been) easily destroyed by mobile fishing gear, have a high associated biodiversity, act as a nursery for some species and, once destroyed, are unlikely to come back.

The identification of Horse Mussel beds as a 'Protected Habitat' by the OSPAR Commission on the Protection of the Marine Environment of the NE Atlantic, as a 'Reef' in the EC Habitats Directive and as a 'Biodiversity Action Plan Habitat' (now variously re-named a 'Principal Marine Habitat', a 'Priority Marine Feature', or a 'Feature of Conservation Importance') means that they are protected as designated features in MPAs and are subject to surveys which have greatly informed our knowledge of their character.

Typically, the beds are found on circalittoral shelly sediments where there is little wave action but moderate tidal flow. Extensive surveys have been carried out in Wales and Scotland. One such survey, at North Cava in Orkney, revealed a density of 80–100 mussels per square metre. The Horse Mussels provide a habitat for many other species that conspicuously include Queen Scallops, brittlestars, sea urchins, hydroids, sea squirts, tube worms, whelks and barnacles. In situ surveys, and collecting clumps of Horse Mussels have revealed a variable number of associated species – from up to 79 at the North Cava site to 300 off Little Ness in the Isle of Man.

A Horse Mussel bed at Burgastoo in Busta Voe, Shetland. Image width *c.* 80cm. '*Modiolus modiolus* beds with fine hydroids and large solitary ascidians on very sheltered circalittoral mixed substrata' (A5.623 / SS.SBR.SMus.ModHAs).

A bed of sparse Horse Mussels *Modiolus modiolus* in Port a' Mheirlich (Conservation Bay), Loch Carron, where strong tidal streams occur. In this image, associated species include the Green Sea Urchin *Psammechinus miliaris* and three species of brittlestars: Common Brittlestar *Ophiothrix fragilis*, Black Brittlestar *Ophiocomina nigra* and Crevice Brittlestar *Ophiopholis aculeata*. Image width *c.* 20cm. 'Modiolus modiolus beds with hydroids and red seaweeds on tide-swept circalittoral mixed substrata' (A5.621 / SS.SBR.SMus.ModT).

A Horse Mussel bed in sheltered conditions in upper Loch Creran. Tube worms *Pomatoceros* sp., Queen Scallop *Aequipecten opercularis*, Common Brittlestar *Ophiothrix fragilis* and the solitary sea squirt *Pyura microcosmus* are particularly conspicuous amongst the *Modiolus*. Image width *c.* 12cm. 'Modiolus modiolus beds with fine hydroids and large solitary ascidians on very sheltered circalittoral mixed substrata' (A5.623 / SS.SBR.SMus.ModHAs).

Maerl beds

Maerl is the name applied to calcified seaweed that forms detached nodules. In areas of moderate tidal flow and shelter from wave action, maerl species can form extensive beds of loosely packed nodules that provide a habitat for a rich variety of species: over 150 macroalgal species and 500 benthic faunal species have been recorded.

Maerl is very slow growing – about 1mm a year – and the beds will have taken many decades, if not centuries, to establish. The age of some beds has been found to be more than 6,000 years. Beds are very susceptible to physical disturbance such as dredging for scallops or for creating navigation channels. In their studies of maerl in Scotland, Jason Hall-Spencer and Geoff Moore observed in 2000: 'Maerl is a 'living sediment'; it is slow to recover from disturbance by towed gear due to infrequent recruitment and extremely slow growth rates'. 'Slow' may be at least several human generations.

Maerl biotopes are included within the EC Habitats Directive as part of 'Sandbanks slightly covered by seawater all of the time'.

Three species of maerl that occur most commonly around Britain. The image width is 5cm for the upper two images and 10cm for the lower image. (A fourth species, *Phymatolithon lusitanicum*, was recorded for the first time in Britain in 2015 and a fifth species, *Lithothamnion erinaceum*, was identified as a species new to science in 2017.)

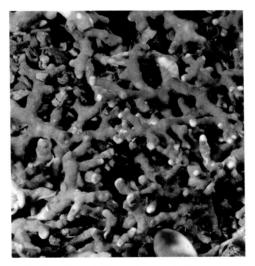

Celtic Maerl *Phymatolithon calcareum* in Strome Narrows, Loch Carron.

Northern Maerl *Lithothamnion glaciale* in Caol Scotnish, Loch Sween.

Southern Maerl *Lithothamnion corallioides* in the Helford River, south Cornwall.

Above: a patch of dense maerl, *Phymatolithon calcareum*, in Strome Narrows, Loch Carron, in December 2000. A Common Sunstar *Crossaster papposus*, Dahlia Anemones *Urticina felina* and a Queen Scallop *Aequipecten opercularis* can also be clearly seen. Image width *c.* 45cm. '*Phymatolithon calcareum* maerl beds in infralittoral clean gravel or coarse sand' (A5.511 / SS.SMp.Mrl. Pcal).

Recording species on the maerl bed off St Mawes in Falmouth Harbour.

The St Mawes maerl bed in Falmouth Harbour with a large Spiny Starfish *Marthasterias glacialis* and a variety of foliose algae. Image width *c.* 60cm. '*Phymatolithon calcareum* maerl beds in infralittoral clean gravel or coarse sand' (A5.511 / SS.SMp.Mrl.Pcal).

File Shell reefs

Beds of the Gaping File Shell *Limaria hians* (also called 'Flame Shells' by divers because of the mass of orange tentacles extending from a red body) are almost entirely hidden from view but are an important structural feature in coarse sediment seabeds. Although occurring along much of the western seaboard of Britain, File Shells only seem to form spectacular beds in Scotland. The bivalve mollusc secretes byssal threads to construct nests that can join together to form reefs that cover several hectares of seabed. The presence of such beds has been known since the 19th century, but exploring their character awaited the use of scuba diving equipment. In the journal *Aquatic Conservation* in 2000, Jason Hall-Spencer and Geoff Moore described how the nests supported a high diversity of associated organisms, including 19 species of algae and 265 species of invertebrates in six discrete nests studied in Loch Fyne. The nests could be very dense and hold more than 700 File Shells per square metre.

The extent of File Shell occurrence and their nests had seemed, in 2000, to be in decline, but surveys since then have suggested a significant increase in their extent and the number of locations where they occur. That increase has a dark side. Maerl beds, in Strome Narrows (Loch Carron) at least, have been displaced by the increased abundance of File Shells, which have consolidated sediments so that algae could grow and smother the maerl. Both maerl beds and File Shell beds are 'Priority Marine Habitats' in Scotland – and one has now reduced the extent of the other.

Gravelly sediment in shallow depths is normally only sparsely colonised by seaweed. The reason why the sediment was visually dominated by Sea Oak *Phycodrys rubens* only became apparent when it was realised that the sediment had become held together by nests produced by Gaping File Shells *Limaria hians*. The two lower pictures are the same area of seabed with the nests pulled back in the second image. Image width: *c.* 1m in the foreground (upper image) and 20cm (lower images). '*Limaria hians* beds in tide-swept sublittoral muddy mixed sediment' (A5.434 / SS.SMx.IMx.Lim).

The Gaping File Shell *Limaria hians* was described as 'The most beautiful of British bivalves' by C. Maurice Yonge and Tom Thompson in the book *Living Marine Molluscs* in 1976. It is the key structural species and the characterising species of the biotope '*Limaria hians* beds in tide-swept sublittoral muddy mixed sediment'. Image width *c.* 4cm.

Serpulid worm reefs

The tube worm *Serpula vermicularis* is widely distributed in British waters, but it very rarely forms reefs. Where it does, those three-dimensional structures provide protection for a large number of mobile species and an attachment point for sessile organisms such as sea squirts, sea firs, bivalve molluscs and other polychaete worms. Species such as hydroids, that attach to the tubes, and brittlestars, that live in amongst the tubes, can raise their feeding structures into stronger currents and out of silty habitats. These castle-like constructions may be easily damaged in storms, but also by divers and photographers who need to take care in their vicinity. The images included here were taken in Loch Creran in 2007: when I returned to the same site in 2016, the condition of the reefs was much poorer.

Serpulid worm reefs constitute a biogenic reef habitat included within the Habitat 'Reefs' of the EC Habitats Directive.

Serpulid worm reefs in Loch Creran. Image width *c.* 1.2m (upper image) and *c.* 15cm (lower image). '*Serpula vermicularis* reefs on very sheltered circalittoral muddy sand' (A5.613 / SS.SBR.PoR.Ser).

Other biogenic reef habitats

Biogenic reefs have been recognised as fragile, easily damaged and often unlikely to return (within several human generations) if destroyed. Horse Mussel beds, maerl beds and serpulid worm reefs all fall into the 'unlikely to return if destroyed' category and are important to protect. But there is a dilemma in conservation: not all biogenic reefs are of conservation importance, but all get lumped into that category despite scientific evidence that they do not all require management action. One particular example that does not need protection is the reef-forming tube worm *Sabellaria spinulosa*. It is a fast growing, opportunistic species that colonises disturbed substrata and is often found in highly turbid or polluted situations, i.e. it benefits from disturbance.

Host species

Some species, whether they like it or not, are habitats for animals that live solely or almost only on that host. These are not always parasites or predators. Sometimes there may be a mutual advantage, for instance when an accompanying anemone provides protection via its stinging cells and gains scraps of food from its hermit crab host (a mutualism). Sometimes, there is no obvious two-way relationship and the attached or associated species is just using the body or body cavity of another species as a way of raising itself off the seabed or obtaining shelter: the host organism neither benefits nor is harmed (known as a commensal or an inquilinistic relationship, sometimes described as being like a tenancy).

Many species prey on others. In this section, I only mention prey relationships for two species that live on Pink Sea Fans but there are many species, especially various families of worms and species of copepod crustaceans, that feed on larger organisms without killing them. Those larger organisms might, therefore, be considered a habitat for the predator or parasite.

Species associated with Pink Sea Fans *Eunicella verrucosa*. The Pink Sea Fan *Eunicella verrucosa* is a horny coral that is common in parts of south-west England. It occurs, in Britain, as far east as Poole Bay on the southern coast of England and as far north as north Pembrokeshire in west Wales. The sea fans attract a variety

A healthy Pink Sea Fan *Eunicella verrucosa*, at Hilsea Point Rock east of Plymouth. Image width *c*. 60cm.

of associated species, some of which are only or mainly found on them. Pink Sea Fans in good health are colonised by the sea slug *Tritonia nilsodhneri* and by the false cowrie *Simnia hiscocki*, both of which seem to live and feed only on the sea fans. These predators seem not to cause significant damage. However, the Sea Fan Anemone *Amphianthus dohrnii*, whilst seemingly using the sea fan only as an attachment, harms the sea fan by smothering tissue and leaving behind bare skeleton, which in turn becomes fouled. Those fouling organisms include the barnacle *Hesperibalanus fallax*, the hydroid *Antennella secundaria* and the sea squirt *Pycnoclavella aurilucens* which, although also living on other species, especially favour Pink Sea Fans.

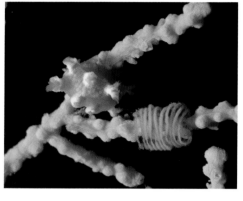

Left: the sea slug *Tritonia nilsohdneri*, which lives exclusively on sea fans. The picture also shows its eggs. Image width *c.* 25mm.

Right: the false cowrie *Simnia hiscocki* was noticed as distinctly different to *Simnia patula* (which lives on sea fingers *Alcyonium* spp.) and was described as new to science in 2011. Image width *c.* 15mm.

Left: the Sea Fan Anemone *Amphianthus dohrnii* uses the Pink Sea Fan as a platform from which to feed on passing plankton, etc. but damages the living tissue through smothering. Image width *c.* 30mm.

Right: the warm-water barnacle *Hesperibalanus fallax* attaches to organic material, including the skeleton of sea fans, and to plastics. It thrived after a disease event in the early 2000s that damaged or killed many Pink Sea Fans. Image width *c.* 25mm.

Left: the sea fir *Antennella secundaria* which grows on organic surfaces including sponges and parchment worm tubes, but especially on sea fans. Image width *c.* 60mm.

Right: the sea squirt *Pycnoclavella aurilucens*, which is found on a variety of substrata but favours dead sea fan branches. Image width *c.* 50mm.

Other host species

Many species act as a host for species that either rely on that species (and are only found there) or are especially found in association with them. Six examples are illustrated below.

The barnacle *Adna anglica* on stony corals. There are five species of stony corals that occur on reef habitats in shallow seas around Britain and all of them may provide a habitat for *Adna anglica*, which is only found on corals and only in the south-west of Britain. The barnacle is most often seen on the Devonshire Cup Coral *Caryophyllia smithii*. The advantages of the relationship seem to be entirely with the barnacle and they may interfere with growth of the coral.

Photographed at Smallmouth on the north Devon coast. Image width *c.* 3cm.

The hydroid *Hydractinia echinata* (Snail Fur) on hermit crabs, *Pagurus bernhardus*. This is a complex hydroid with three different types of polyps in addition to the male and female reproductive polyps. The hermit crab most likely benefits from protection afforded by the stinging cells of the hydroid and the hydroid from scraps of food as the crab eats.

Photographed in the Helford River, south Cornwall. Image width c. 4cm.

Photographed in Plymouth Sound. Image width *c.* 7cm.

Cloak Anemone *Adamsia palliata* encircling hermit crabs *Pagurus prideaux*. Another coelenterate, this time a sea anemone, that has an obligatory relationship with a hermit crab, *Pagurus prideaux*. In this case, it seems that the hermit crab benefits from the relationship in that the anemone creates a chitinous shell that carries on growing and means that, as the crab grows, it does not need to seek out larger shells. The anemone very readily emits defensive pink threads (acontia) if disturbed. The hermit crabs, photographed in mid-December, are a pre-mating couple.

The Parasitic Anemone *Calliactis parasitica* on hermit crab shells and other organic substrata. It is unfortunate that the mis-named Parasitic Anemone has retained that name, as it merely attaches to, usually, shells of Common Whelks *Buccinum undatum* occupied by hermit crabs and sometimes to the claws of Spiny Spider Crabs *Maja brachydactyla*.

Photographed off West Hoe, Plymouth Sound. Image width c. 7cm.

Photographed in Plymouth Sound. Image width c. 10cm.

The Leach's Spider Crab *Inachus phalangium* and the Snakelocks Anemone *Anemonia viridis*. During summer months especially, these spider crabs can be seen sheltering under Snakelocks Anemones and often crawling amongst the tentacles. The anemone may be used just as shelter and protection (the anemone's tentacles have powerful stinging cells) with no advantage for the anemone, although the crab may perform some cleaning function. There is a similar relationship between the Anemone Shrimp *Periclimenes sagittifer* and Snakelocks Anemones (illustrated in the next chapter: 'Change' (page 227)).

The bivalve mollusc *Montacuta substriata* and burrowing sea urchins.
Sea urchins and their burrows are an attractive habitat for this filter-feeding bivalve mollusc which gathers around the area where waste (which is highly nutritious) is disposed of. The bivalves are characteristically found at or near the urchin's anus with the smaller ones especially attached by byssus threads to the spines of the sea urchin. The relationship is considered commensal, with the mollusc obtaining benefit and doing no harm to the urchin. Photographed here, the burrowing Purple Heart Urchin *Spatangus purpureus*, removed from the sediment and turned upside down to reveal the bivalve mollusc *Montacuta substriata* attached to the anal spines.

Photographed in Wembury Bay, south Devon. Image width *c.* 10cm.

Change

Expect change in seabed communities: especially seasonal change, but also fluctuations from year to year, at a specific location and within geographical areas. However, those changes are likely to be what scientists call 'noise' in the system and, in the long term, don't expect too much change. Philip Henry Gosse, writing in his *Actinologia Britannica* in 1860 was surprised to observe of a rockpool in Torbay that he had visited six years previously: 'The shore was as if I had left it but yesterday'. Going back to that rockpool in 2010, most of the species that Gosse had described were, I found, still there 150 years later. Similarly, the subtidal area in Torbay that was mentioned in the first chapter was populated mostly by the same species as had been described 150 years previously and, in 2010, what are essentially subtidal communities in the caves at St Catherine's Island near Tenby were almost identical to how Gosse described them in the 1850s.

Environmental conditions at a particular location such as bottom type, exposure to wave action and tidal currents, depth, salinity and turbidity ranges over the course of a year are unlikely to change, so that the particular seabed communities that develop as a result of those conditions will most likely be the same from year to year. However, the species that characterise those communities may be subject to change as there are good and poor years for recruitment, there are disease events and there is occasional severe weather that may affect the abundance of component species. John Gray, in a paper published in 1977, illustrated types of stability in seabed ecosystems as a ball in a cup: disturb the cup and the ball (the community) moves about (the species composition varies) but the ball stays in the cup (the community persists); disturb the cup greatly and the ball may be displaced either to a state where the community becomes extinct or, more likely, changes into a different community. Most communities have a high degree of long-term stability but vary slightly with time and season.

Many seabed marine biological features persist on decadal – if not century – timescales because the environmental conditions that determine the habitat and which species will thrive there have been unchanging. However, new recruits to the study of marine biology may take too much notice of the sorts of change that can occur on land. There, maintaining the status quo is often a matter of manipulating nature to preserve or create what is desired to support richness and diversity or even perceived beauty. If habitats are not grazed, flooded, coppiced, re-seeded, etc., they change, and old data may not accurately indicate what is now present at a location. In the sea, many habitats are close to natural and persist, with their associated biota, with time. Now that we are getting quite good at providing access to data and information (see the final chapter: 'Technology takes off'), perhaps the plea especially to those involved in marine environmental protection and management should be, 'Don't ignore old data'.

As on the land, there are perennial species, annual species, long-lived and short-lived species within a particular habitat. The degree of persistence of seabed communities at a location may depend on the longevity of component species. Many are very long-lived, slow-growing and recruit close to their parents. A prime example is the Yellow

Staghorn Sponge *Axinella dissimilis*, a species frequently found on some reef habitats and illustrated here. Amongst the algae, species of maerl (calcified seaweed that forms twig-like growths) are also very slow-growing species – less than about 1mm a year in length. Beds of maerl have a rich associated community but appear to have very specialised requirements for growth. The only live beds known in England are in the Fal and adjacent Helford Rivers, where they have been known since the mid-1960s but have doubtless been there much longer. Radiocarbon dating has revealed that some maerl beds have, as already mentioned, been in the same place for 6,000 years. Knowing the life history traits (growth rates, longevity, dispersal characteristics, etc.) of the species that constitute a particular biotope is an essential part of identifying the likelihood of that biotope being damaged by factors such as physical disturbance, and whether or not it will return if lost.

The records that I and many others have collected in over 40 years of marine biological survey for purposes of environmental protection and management provide important reference points for identifying the degree and character of change. I continue to investigate many of the reef habitats at locations I first dived in the late 1960s and the assemblages of species there are remarkably constant. Again, there are

Some species that characterise seabed communities may be exceptionally long-lived and slow-growing, persisting through conditions that are adverse for other species. The Yellow Staghorn Sponge *Axinella dissimilis* has been shown to grow extremely slowly (less than 1mm a year) and may be 300mm tall. They have continued to be present at Lundy whilst some other species have declined. Image width *c.* 12cm.

In 2007, scientists from the Marine Biological Association resampled the sites that E. J. Allen had sampled at the end of the 19th century (see 'Exploration and discovery: the first 150 years'). In the paper published by Evelina Capasso and colleagues in the *Journal of the Marine Biological Association*, the authors concluded, 'the differences observed are concomitant with changes generally associated with disturbance from demersal fishing activities'. Image: Hilmar Hinz.

exceptions, including declines in the abundance of some species and increases in the abundance of others. Some of those changes may be due to long-term fluctuations in environmental conditions and even regime shifts such as that identified as having occurred in the North Sea in the mid-1980s, or which are a consequence of the sort of decadal-scale changes that characterise what has become known as the 'Russell Cycle' in the English Channel, described on page 221.

Ports, harbours and developed estuaries will most likely have changed in biological character due to construction, dredging and contaminants but also because they are often now colonised by non-native species brought by shipping or via mariculture from all over the world. Some species, such as native oysters *Ostrea edulis* and seagrass *Zostera marina*, have declined enormously from their abundances a hundred years ago. Whilst some of the reasons for decline (such as nutrient enrichment, sedimentation, disease and, in the case of oysters, over-exploitation) are clear, why recovery has not occurred to any significant extent is unclear. Significant changes that can be linked to fishing are known to have occurred on level seabed habitats and on low reefs, although reef habitats are generally no-go areas for heavy mobile fishing gear.

Change in the species present and in the abundance of species in seabed habitats is to be expected as a result of seawater warming and, possibly, ocean acidification. A few instances are noted in this chapter.

SEASONS IN THE SEA

We are all familiar with the changes that occur in terrestrial wildlife and in garden plants through the year, and there are similar seasonal changes on and within the seabed. In the sea, spring is a time for animals to reproduce and for new growth to occur whilst, in autumn, animals and plants mostly shut up shop ready for the winter. The plants and animals colonising seabed habitats may also change in character, and even in respect of which species are dominant through the year. Some species are present – or present in large numbers – only at certain times of the year. As on the land, those changes may not occur all at the same time (although mid to late winter and autumn are important times for change in the sea). The changes described here are to give examples of what occurs, without in any way attempting to be comprehensive. Whatever the differences in the abundance of seabed species at different times of year (and there may be none of significance in some habitats), the researcher undertaking surveys, and especially annual monitoring studies, needs to be aware that seasonal change occurs.

The following examples and images of seasonal changes in seabed species are from near my home base in Plymouth. It is also from Plymouth that much of our knowledge of the time of reproduction and settlement of animal species, at least, has been recorded. The *Plymouth Marine Fauna* (last published in 1957 and now available online) is a treasure trove of information about the biology of seabed species. Now, the Marine Life Information Network (MarLIN) website brings together such information as part of the catalogue of biological traits (published separately as Biotic) that help us to understand the life history of species and much else.

Left: the aft deck of ex-HMS *Scylla* in mid-July. There is summer growth of red algae (especially thread weed *Vertebrata byssoides* and Netted Wing Weed *Dictyopteris polypodioides*) and of antenna hydroids *Nemertesia antennina*. Image width *c.* 1m.

Right: the aft deck of ex-HMS *Scylla* in late January. The deck is largely bare except for tube worms and (bottom right in the image) fresh growth of Sea Beech *Delesseria sanguinea*. Image width *c.* 1m.

In March and April, Cuvie *Laminaria hyperborea* are shedding their old fronds and new ones are growing below the pinch. Photographed at Hand Deeps south-west of Plymouth. Image width *c.* 2m.

Below: detail showing the pinch (centre) between old and new fronds. Photographed in Wembury Bay in mid-April. Image width *c.* 60cm.

Spring

Whilst many fish are still tucked away in fissures and crevices, the underwater gardens will be blooming. Clean, fresh seaweed growth (of both perennial and annual algae) will be showing itself and, of the attached animal species, it is the hydroids that are especially spectacular. There will be new growth of antenna hydroids – soon to become like fields of grass in the underwater landscape. The bright red polyps of tubular hydroids will appear especially where strong tidal currents bring food. Those bright red polyps may be short-lived if it is a good year for the sea slugs that feed on them.

Spring starts early

For many algae, spring comes early; as soon as the nights start to get shorter in length, they will begin to grow. For some animal species, winter closure also finishes early, perhaps in readiness for the plankton bloom in April and May. Species such as Dead Men's Fingers *Alcyonium digitatum* and Plumose Anemones *Metridium dianthus* begin to expand rather than to be contracted. Often, the most spectacular shows of Jewel Anemones *Corynactis viridis* are in what would be considered winter (January and February) on land.

In this image, new growth of the perennial Sea Beech *Delesseria sanguinea* can be seen on the rails of ex-HMS *Scylla* in early February. The midribs of the previous year's growth are still present. Dead Men's Fingers are fully expanded, in contrast to their closed period from about October through to December. Image width *c.* 30cm.

Many species spawn in the spring, creating larvae that might spend weeks or even months in the water column, dispersing to new locations.

Echinoderms may be seen spawning in spring. **Top left:** an Edible Sea Urchin *Echinus esculentus* is spawning in mid-March (image width *c.* 10cm). **Bottom left:** a Spiny Starfish *Marthasterias glacialis* has raised itself into the current to release what are most likely eggs (the orange flush) in mid-April (image width *c.* 25cm) – other spiny starfish a few metres away were producing sperm or eggs at the same time. **Right:** the Cotton Spinner *Holothuria forskali* sea cucumber has climbed (with many others) to a high point on its reef in late April and is ready to spawn (image width *c.* 25cm).

Early May, and growths of the athecate (naked polyp) hydroids *Garveia nutans* and *Ectopleura larynx* have blossomed and are reproducing in a tide-swept habitat at the entrance to the Tamar River in Plymouth Sound. Image width *c.* 6cm.

Spectacularly colourful growths of the tubular hydroid *Tubularia indivisa* colonise rocks and wrecks in late April and early May, but will likely be found soon afterwards by the sea slugs (nudibranchs) that feed on them and lay their eggs on the now-dead stalks. The sea slug is *Fjordia lineata*. Image widths *c.* 12cm for hydroid colonies and *c.* 4cm for the sea slug.

Lobsters *Homarus gammarus* are said by fishermen in some parts of Britain to 'start to walk' once the seawater temperature rises above 12°C. This means that pots and creels will start to make significant catches. Image width *c.* 40cm.

Spring is a busy time for breeding and one of the much-anticipated events amongst scuba divers is the arrival of cuttlefish inshore to court, mate and lay their eggs from around the end of April into May. Fish are much less conspicuous in their breeding habits but, by the end of April, male Corkwing Wrasse *Symphodus melops* will be building nests out of seaweed on suitable rock shelves and Tompot Blennies *Parablennius gattorugine* will be guarding their eggs under boulders or in fissures.

A female preparing to lay eggs, accompanied by two male Common Cuttlefish *Sepia officinalis* in mid-May in Torbay. Image width *c.* 50cm. The characteristic grape-like eggs are attached to both natural surfaces such as Pod Weed *Halidrys siliquosa* but, in this image, to a plastic tie on a submerged cable in Plymouth Sound, and much earlier than expected: on 9 April 2017. Image width *c.* 12cm.

A male Tompot Blenny *Parablennius gattorugine* guarding eggs in a horizontal fissure, photographed in late May 2017. Image width *c.* 15cm.

Summer

A male Corkwing Wrasse *Symphodus melops* tends its nest. The outer surface is of tough coralline algae and the inner part is of soft algae. In deeper water, the outer part may be of branching sea mats. Nest building begins in April and the nest or nests are maintained by the male through into August. Image width *c.* 50cm.

The Rosy Featherstar *Antedon bifida* produces short-lived pelagic larvae that settle as miniature 'sea lilies' attached by a stalk to their chosen surface. The top images are taken in early June and show a group of young attached featherstars (**left**), and (**right**) showing clearly the sea lily-like stalk. The newly settled individuals grow rapidly and soon become small featherstars (**lower image left** in early August: image width *c.* 3cm). The young and fully grown featherstars attach to the seabed with their claw-like cirri (**lower image right**, in late September: image width *c.* 10cm).

Grey Triggerfish *Balistes capriscus* arrive as early as mid-May on western coasts – but not every year. It seems that they migrate to British shores (assisted by oceanic currents) perhaps from across the Atlantic, or perhaps from further south in the Eastern Atlantic. However, they are a warm water species and seem not to make the return journey, but become comatose with cold and are washed up dead on strandlines in December and January. This image was taken in mid-August. Image width *c.* 35cm.

For many of the species that have produced larvae in spring, those larvae have spent their time in the plankton (if they have a planktonic larva) and have now settled onto the seabed. Some with long-lived larvae may still be settling and the echinoderms are especially likely to spend a long period in the plankton.

The larvae of seabed species are rarely visible to the naked eye but, by late summer, those of the genus *Luidia* are miniature adults and, surrounded by a gelatinous sheath, can often be seen in their thousands drifting past a decompressing diver. In the image here, the larva is accompanied by an amphipod crustacean, most likely *Hyperia galba* (image width *c.* 15mm). The larva was photographed in the Isles of Scilly in early October. The newly settled *Luidia ciliaris* are shown photographed in early September at Lundy (Image width *c.* 15cm). The fully grown adult, which preys on other echinoderms, is here about 50cm across, photographed at Hand Deeps south-west of Plymouth.

By mid-August, these kelp plants, Furbelows *Saccorhiza polyschides,* on one of the Eddystone reefs are covered in a summer growth of encrusting sea mats, greatly reducing their ability to photosynthesise. Image width *c.* 2m.

By the end of the summer, success for some species means problems for others. Algae are greatly covered by encrusting species that have settled during the summer, and it is time to shed fronds – or perhaps those damaged fronds just break up with wave action.

From late summer onwards, enigmatic gatherings of Spiny Spider Crabs *Maja brachydactyla* occur. Their purpose seems to be to moult; there is protection in numbers, but many die. Thousands may occur together in mounds.

A mound of moulting Spiny Spider Crabs *Maja brachydactyla* at Babbacombe in Torbay, photographed in early September. Image width *c.* 1.5m.

Inside the bridge on ex-HMS *Scylla* in mid-July 2013. Exuberant life with Plumose Anemones, soft corals (Dead Men's Fingers), hydroids and much else.

Inside the bridge on ex-HMS *Scylla* in early November 2015. Plumose Anemones and soft corals (Dead Men's Fingers) are closed up and summer species (including most divers) have disappeared.

Autumn

A period of shut-down, migration, of loss due to predation or natural end-of-life which may start as early as September in some species. Even before seawater temperature starts to fall after the end of September, some species will become inactive. For instance, Richard Hartnoll, working at the Port Erin Marine Laboratory, studied the rhythm of feeding in Dead Men's Fingers *Alcyonium digitatum*. From January through to about September, colonies expanded and contracted their polyps several times during the day. From about October through to December, the colonies remained contracted and did not feed. Often, the surface of the colonies

An expanded colony of Dead Men's Fingers *Alcyonium digitatum* at Hand Deeps south-west of Plymouth in mid-September. Image width *c.* 12cm.

A colony of Dead Men's Fingers *Alcyonium digitatum* at Hand Deeps in late-October. The polyps have been withdrawn and the surface has growths of diatom algae and sea firs. Image width *c.* 12cm.

A colony of the Football Sea Squirt *Diazona violacea* photographed off Plymouth Sound in mid-April. The zooids are now well-established.

A colony of the Football Sea Squirt *Diazona violacea* photographed off Plymouth Sound in early December. The zooids that were permanently expanded during late winter through to autumn have been shed but there are signs of new ones growing below the surface in this colony.

became covered in a layer of species such as diatom algae and sea firs. The season of prolonged inactivity coincided with the final months of gonad maturation and the shedding of the colonising species and re-expansion of polyps immediately preceded spawning. In the Football Sea Squirt *Diazona violacea* the change is more dramatic, as the feeding zooids of the larger colonies, at least, are shed in about October and new ones start to grow in about December.

By early November, seawater temperatures are just beginning to drop significantly; the autumn plankton bloom is coming to an end; annual seaweeds are being or have been swept away by storms; some perennial seaweeds are losing their fronds (eaten or broken up); some sea anemones and soft corals have become inactive until mid or late winter, and many sea firs are just dead stalks. But, some species are showing the success of summer reproductive effort, and wrasse numbers in particular are likely to be high on shallow reefs before they seek shelter during the rest of the winter.

In early November, seawater temperatures are just beginning to fall significantly but summer breeding success still shows itself. Wrasse at the Eddystone reef south of Plymouth in early November 2015.

By the end of summer, some annual species have grown large but, in the case of many algae, are now covered in encrusting sea mats. Here, annual kelps, Furbelows *Saccorhiza polyschides*, are pictured on the Eddystone reefs in mid-September 2015. Image width *c.* 1m.

By early autumn, Furbelows *Saccorhiza polyschides* have been mainly swept away, although the bulbous holdfasts may (as in the image) remain. Perennial kelps, Cuvie *Laminaria hyperborea*, will resist winter storms and now visually dominate the reefs. Eddystone reefs in late October 2015. Image width *c.* 1m.

Winter

By mid-winter (December and January), reefs are largely devoid of annual species and the broken debris of the summer extravaganza have been swept away. Many territorial but resident fish have tucked themselves away into often deep fissures and holes. There are rarely the large numbers of highly mobile fish such as Pollack *Pollachius pollachius* that occur in summer, and any Grey Triggerfish *Balistes capriscus* that found their way to Britain in the summer will be dead.

Despite the apparent inhospitality of winter, some species will be breeding and spawning or laying their eggs – perhaps in anticipation of a plentiful food supply later in the year.

The hydroid *Gymnangium montagui* produces reproductive bodies (gonozooids) along the central stalk in midwinter – here it is seen in mid-December. Image width *c.* 10mm.

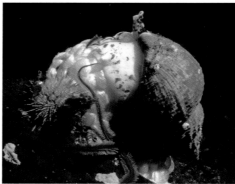

The Common Whelk *Buccinum undatum* laying eggs in Plymouth Sound on 22 December 2008 (left) and in Loch Carron on 21 December 2011 (right). Egg laying continues for another couple of months and the young hatch as tiny crawl-away miniature adults in spring. Image widths *c.* 12cm.

Algae often show marked changes in abundance and condition over the course of a year. Some are perennial and are there all the time, including the kelp Cuvie *Laminaria hyperborea* and such red algae as Siphoned Feather Weed *Heterosiphonia plumosa*. Others, such as the large kelp Furbelows *Saccorhiza polyschides* grow quickly in spring and early summer through to the autumn (by which time they are usually heavily encrusted with sea mats) and then are swept away on the open coast by winter storms. Some may survive into a second year, especially in sheltered conditions. Sea Beech *Delesseria sanguinea* is a perennial red alga that dies back to stalks in the autumn and starts to show the growth of new blades in about January.

The reasons for seasonal change are many, and are different for different species. In a paper entitled *The seasons in the subtidal*, Joanna Jones (Kain), who undertook work in the Isle of Man and Firth of Clyde, concludes, 'There is clearly a mixture of factors causing seasonal change in subtidal algae; sometimes one is important and sometimes another. There is no evidence for temperature having particular importance but there is evidence for light being a frequent cause of change.' Similar conclusions, but referring to different environmental factors, could be applied to many species that we know about. There are many more species for which we know little or nothing of growth, seasonality or persistence.

HUMAN ACTIVITIES

Inevitably, human activities have greatly changed some areas of seabed, especially in inshore areas. The impacts of mobile fishing gear on seabed habitats and species is at the forefront of activities causing widespread change. Some localised changes are mentioned in the section on 'Wrecks and other artificial substrata'. Other changes may occur as a result of dredging in ports and the subsequent disposal of dredged sediment, as well as end-of-the-pipe pollution effects. More widespread changes are likely as a result of diffuse pollution, but are difficult to pin down. The transfer of species from different regions of the world causes conspicuous change to seabed habitats and is included in brief on pages 226 and 227. However, this book focuses primarily on habitats that are as close as possible to their natural state.

DECADAL SCALE CHANGE

There are rhythms in the sea and there are expected events that occur sporadically but their timing is unpredictable (stochastic change). Such events may reset the character of seabed communities or abundance of a species. Long-term fluctuations are difficult to describe as cycles because, for most species and communities, data has not been collected for long enough to be sure that there is a repeated period over which the change is occurring.

Trying to understand what determines fluctuations in the abundance of species or changes in communities, let alone predict them, has challenged researchers since at least the 1920s. Most of their studies have concentrated on the plankton and on water column chemistry. During those studies, it became clear that there were intricate balances among nutrients, plants and animals that varied every year. But there were also possible decadal scale changes with species reducing in abundance after a boom and that boom not being repeated for 30 or 40 years.

The 'Russell Cycle'

In their 1976 paper on *The biological response in the sea to climatic changes*, D. H. Cushing and R. R. Dickson recorded variations in the chemical conditions and the plankton out of Plymouth, the causes of which 'must have penetrated all the nooks and crannies of the ecosystem' and resulted in 'consequential changes involving the benthos'. The changes they especially recorded were in phosphorus, macroplankton and summer spawners, which were reduced in 1930/31 and recovered 40 years later in 1970/71. The spring spawners declined in 1935, five years after the initial change and they recovered in 1965, five years before the reversal. Such changes in the English Channel have been named the 'Russell Cycle' after Sir Frederick Russell who worked at and was Director of the Marine Biological Association laboratory in Plymouth.

Sir Frederick Stratten Russell FRS in the on-board laboratory on RV *Salpa* in 1937.

Standard plankton hauls reveal years in which plankton (mainly copepod Crustacea and pilchard eggs here) are abundant and many years in between where they are scarce, creating feast and famine for species that feed on plankton. Image: Alan Southward.

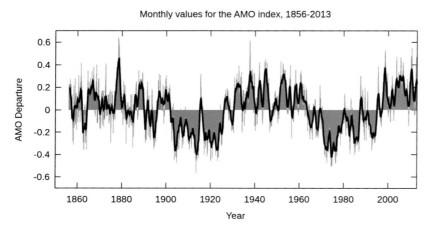

Monthly values for the AMO index, 1856-2013

A possible driver of change in seabed communities – the Atlantic Multidecadal Oscillation. The 'AMO Departure' refers to degrees centigrade from the mean. Source: NOAA.

Decadal scale changes in climate that have possible implications for marine life are now seen as oscillations rather than cycles. The North Atlantic Oscillation (NAO) controls the strength and direction of westerly winds and location of storm tracks across the North Atlantic and varies over time with no particular periodicity. The Atlantic Multidecadal Oscillation (AMO) is a climate cycle that affects the sea surface temperature of the North Atlantic Ocean on multidecadal timescales, with cold phases from approximately 1900–1925 and 1971–1994 and warm phases from 1875–1899, 1926–1970 and 1990 to present. Making any link between such oscillations in climate and changes in seabed marine life requires a much longer time-series of observations of species abundances and composition of seabed communities than we have at present.

Declines and disappearances or 'boom and bust' situations are being seen in many seabed species, although finding patterns awaits more observations (and those observations being recorded!). Some examples for seabed species are given next.

Until about 2005, the Sea Fan Anemone *Amphianthus dohrnii* was abundant on the large aggregations of Pink Sea Fans *Eunicella verrucosa* on the wreck of the *Persier* in Bigbury Bay, east of Plymouth. Numbers declined and, after about 2014, none could be found there. Further west, in a similar situation on the wreck of the SS *Rosehill*, numbers appear to have increased since the early 2000s. These are most likely natural fluctuations. Image width *c*. 2cm.

European Spiny Lobster *Palinurus elephas*
– localised extinction and recovery (but after nearly 40 years)

European Spiny Lobsters, also known as Crawfish, are likely to be key functional species in seabed ecosystems: preying on molluscs and other crustaceans as well as scavenging. Large individuals can have a carapace length of up to 25cm. They shelter in cavities on rocky seabeds, where they are often long-term residents. This individual was photographed in the Firth of Lorn in 2004. The left antenna was distinctive and the individual had been seen in the same location for over three years before disappearing after netting for Spiny Lobsters had taken place. Image width *c.* 60cm.

They were delicious, but overfishing with nets and by diving in the late 1960s and early 1970s depleted numbers enormously. By the mid-1970s this spectacular shellfish had become extinct in some locations and enormously reduced in others where it had previously been common. The author is pictured returning from a shore dive in the Isles of Scilly in 1969.

Larvae are shed between about March and June and spend, it is thought, about 150 days in open water before settling as individuals of about 2cm body length in late summer. The larvae will be carried by oceanic currents, and recruitment to south-west Britain may be from Brittany and perhaps especially when sporadic strong currents occur. Larvae may also hitch-hike on jellyfish on which they feed. Image width *c.* 25mm. Image: Richard Kirby.

A few recruits (small individuals) have been seen since 2007 in south-west England and Wales but recruitment at many locations around Britain took off in about 2014. By 2017, at some locations in south-west England as many as six (some divers claim 20) may be seen on a dive where none had been seen for 40 years. Left: European Spiny Lobster recruit in one of the boiler tubes on the wreck of the *Persier*, east of Plymouth in August 2015. Image width *c.* 25cm.

The nationally rare Sunset Cup Coral *Leptopsammia pruvoti* is at the northern recorded limit of its distribution at Lundy. Until the mid-1980s, numbers were high in discrete colonies at three locations there. By 2015, the populations had fallen to about one-third of those censused in 1984 and one colony had entirely disappeared. Successful reproduction and settlement seems not to be keeping up with mortality. Populations off Plymouth appear to be thriving. Wide fluctuations may be expected in the abundance of species at the limits of their range, but possible anthropogenic causes such as high levels of nitrates in the water should be investigated. Image width *c.* 3cm.

The Football Sea Squirt *Diazona violacea* was (within the memory of scuba divers and from surveys in 1986) an unusual species to observe on the reefs out of Plymouth. In 2008, there were large numbers of small colonies observed at several locations and the species has become present in densities of several per square metre in some places. Here, they are seen on the submerged cliffline (the Dropoff) offshore of Plymouth Sound at 30m depth in July 2011. The outburst seems to be entirely natural. Image width *c.* 1m.

Sea slugs (nudibranchs) have been known since the 19th century to go missing for several decades before reappearing: a natural cycle of boom and bust. The spectacularly colourful *Polycera elegans* were commonly seen at Lundy and around Skomer at least until 1986, when they were last seen at Lundy. Look out for their return. Image width *c.* 3cm.

Pink Sea Fans *Eunicella verrucosa* suffered a disease event in the early 2000s in south-west England that killed some and left others partly covered in fouling organisms where dead skeleton had been left exposed. The image is from Lundy in August 2001. Mortality was the result of infection by a *Vibrio* bacterium (determined by Jason Hall-Spencer, James Pike and Colin Munn from Plymouth University). Perhaps not the first time though: this quote is from the *Plymouth Marine Fauna*: 'latter half Aug. and first half of Sept. 1924; Capt Lord reported that a great amount of *Eunicella* brought up was dead; many colonies brought in were partially dead; none in such good condition as in the previous July'. The disease has not been seen since about 2004 but by 2016, Lundy populations had not yet recovered their numbers. Image width *c.* 12cm.

Some seabed species have declined enormously in abundance, but there has been no apparent or substantial recovery. Two examples are seagrass *Zostera marina* and native oysters *Ostrea edulis*, which both declined (probably coincidentally and for different reasons) in the 1930s.

A healthy bed of seagrass *Zostera marina* in Falmouth Harbour. Image width *c.* 50cm.

A solitary native oyster *Ostrea edulis* in the Cattewater at the entrance to the River Plym estuary, photographed in January 2017. There was once a commercial bed here. Image width *c.* 14cm.

NON-NATIVE SPECIES

The distinctive character of the major biogeographical subregions of the world's oceans relies on the presence of barriers to confine species to separate areas. Those barriers include land masses that block or direct the flow of oceanic currents, and waters that are too cold or too warm for a species to survive passage through them as adults, larvae or spores. Those barriers can be breached by vessels that either carry larvae in their ballast water or have organisms fouling on the hull that survive the brief shock of unsuitable temperatures during a rapid passage. They may also be breached as previously ice-locked passages (the Northwest Passage) become ice-free or because of construction of canals that link two oceans. The Suez Canal, and especially the recently completed much larger Suez Canal, is facilitating a massive invasion of Indo-Pacific species from the Red Sea to the Eastern Mediterranean. Around Britain, shipping is a major vector of non-native species and so also is mariculture. Commercially grown species, together with the organisms that hitch-hike on them, are transported from distant parts of the world to grow as food or, for example, production of alginates.

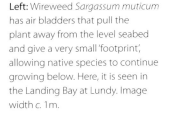

Left: Wireweed *Sargassum muticum* has air bladders that pull the plant away from the level seabed and give a very small 'footprint', allowing native species to continue growing below. Here, it is seen in the Landing Bay at Lundy. Image width *c.* 1m.

Right: Slipper Limpets *Crepidula fornicata* attached to a scallop from Plymouth Sound. Such additions to the shells are unwelcome by fishermen but, more importantly, dense beds of the limpets can change seabed habitat and displace native species. Image width *c.* 15cm.

Left: Wakame *Undaria pinnatifida* is much more successful at attaching to vertical rock surfaces than native kelps. It is pictured at Devil's Point at the entrance to the Tamar River. Image width *c.* 70cm.

Right: the Carpet Sea Squirt *Didemnum vexillum* smothering areas of seabed and the associated species at Herne Bay in Kent in July 2016. Image: Chris Weaver (from a video). Image width *c.* 25cm.

The red alga *Dasysiphonia japonica* has become a dominant feature of some shallow-water areas of sea lochs, displacing native species. Photographed at Loch Carron. Image width *c.* 10cm.

Many non-native species fit in to existing seabed communities and may not significantly change them. Others, such as Wireweed *Sargassum muticum* may appear visually dominant, but research suggests that they do not harm native species unless very abundant. Others, such as Slipper Limpets *Crepidula fornicata*, may become dominant, displacing native species and changing the habitat as their faeces make what had been coarse sediment muddy. They are also an annoyance to scallop fishermen as they commonly attach to scallop shells. Some non-native species, such as the Wakame *Undaria pinnatifida* (a seaweed from south-east Asia imported to Europe for mariculture) may thrive on surfaces that similar native species, other kelps in this case, do not thrive on, and the non-native may displace them. Another seaweed, *Dasysiphonia japonica*, has become common in western Scotland on shallow rocky areas, displacing native species. Perhaps those that have the greatest potential to smother native species and biotopes are sea squirts, the current most worrisome one being the Carpet Sea Squirt *Didemnum vexillum*, which has the potential to smother and kill existing biota living attached to the seabed.

Left: a warm water species, the Variable Blenny *Parablennius pilicornis*, was first recorded in British waters in 2007 by Rob Spray and Dawn Watson and continues to be seen in Plymouth Sound, including laying eggs. Image width *c.* 8cm.

Right: the Anemone Shrimp *Periclimenes sagittifer*, known for many years from the Channel Isles, was first recorded in Britain at Swanage in 2007 by Matt Doggett and Polly Whyte and here, at Babbacombe in Torbay, by Terry Griffiths and Dan Bolt in 2015. Photographed in September 2016.

SEAWATER WARMING

Sea surface temperatures around the British Isles have risen, on average, by about 1.0°C since the mid-1800s, including by about 0.7°C from 1979 to 2009, according to the Hadley Centre for Climate Research. There is some unevenness in those figures and it seems that northern waters, including the North Sea, have seen higher temperature rises than south-western waters. Also, sea surface temperatures are greatly affected by air temperatures and it is seabed (not sea surface) measurements of temperature that are needed to interpret change in seabed communities. Very large changes in the character of planktonic communities and in fish populations have been recorded as a result of seawater warming and there have been range extensions of some intertidal species. Southern or warm water seabed species with long-lived dispersal stages may settle in areas where they can now survive. Warm water species that have very poor dispersal ability may increase in abundance where they currently occur but not extend their range. There is little clear evidence yet of change in subtidal seabed species distribution or abundance that can be ascribed to seawater warming, but some species that have apparently benefitted from seawater warming or have retreated are illustrated in the previous section. There have been some 'new arrivals' of southern species in the past few years. Some of those species are potentially highly mobile and some may have hitched a lift with vessels and some, in the case of *Hesperibalanus fallax* (see image, page 198), on floating plastic debris. Southern species may benefit from warming temperatures, such as Sunset Cup Corals *Leptopsammia pruvoti* (page 224) and Pink Sea Fans *Eunicella verrucosa* (pages 197, 222, 225), but most likely with increased abundance where they are currently found, as neither have good larval dispersal powers. There are other species that are cold-water and northern in their distribution, such as the sea urchin *Strongylocentrotus droebachiensis*, the holothurian *Cucumaria frondosa* and the Wolf Fish *Anarhichas lupus*, that currently persist in the same abundances as always. There are species to especially look out for, as their distribution might change (range extensions and retractions) such as Yarrell's Blenny *Chirolophis ascanii* and the Red Cushion Star *Porania pulvillus*, which are still present in south-west England but are northern in their distribution. From the north, the Northern Sea Fan *Swiftia pallida*, the Northern Stone Crab *Lithodes maja*, the Bottlebrush Hydroid *Thuiaria thuja*, Horse Mussels *Modiolus modiolus* and the Deeplet Sea Anemone *Bolocera tuediae* are further examples of conspicuous northern species seemingly at risk from warming seas. There are potential changes in seabed biology that could result from new interactions. For instance, wrasse are not abundant in Scotland – most likely because temperatures are too low. Wrasse eat Green Sea Urchins *Psammechinus miliaris* (see the description of colonisation on ex-HMS *Scylla*, page 125) and it is Green Sea Urchins that graze rocks in sheltered areas in Scotland to the extent that almost only encrusting coralline algae survive on them in wave-sheltered areas.

Predicting which species may be winners and which losers as seawater temperature changes is no simple matter. Account needs to be taken of hydrographical and geographical barriers to spread and the life history characteristics (reproductive mode, dispersal capability and longevity) of species. The paper that I published with Alan Southward, Ian Tittley and Steve Hawkins in 2004 on the effects of changing temperatures on benthic marine life in Britain and Ireland in the journal *Aquatic Conservation* should dispel any simplistic ideas of likely change.

There are other aspects of climate change that may affect seabed marine life, including sea level rise and increased storminess. Bundled in with climate change is ocean acidification: another factor that may affect some seabed marine life.

WHAT YOU CAN DO

The flavours of change shown in this section of the book are very selective and there are many more that scuba divers in particular will observe. Those sightings can help us to better understand events in the sea through the year, but also their timing from year to year. As climate change progresses, timings of events might get earlier or later in the year and recording the timing of change (phenology) will help us to understand how the underwater world is adjusting. The interested reader can find out much of what we know via the MarLIN and Biotic web sites and even more if they go back to the many scientific papers and regional accounts of species such as the *Plymouth Marine Fauna*. Do contribute your observations via one of the many reporting schemes.

Protecting what we have

The diversity of seabed habitats and associated species present in the shallow seas around Britain is remarkable. Those shallow seas are important for commerce, for recreation, for energy generation, for food supply and for much else. Whether or not seabed habitats and their associated species will be adversely affected by the pressures generated by those activities is important to understand and, if harmful to biodiversity and the benefits we derive from the sea, relevant activities need to be managed to prevent or minimise harm. That management is what conservation is all about.

The drivers for biodiversity conservation range from philosophical (our duty of care for an environment that we can all too easily damage, and for species having a right to exist, whether or not they are of benefit to us) to immensely practical reasons concerned with the services that the sea provides (from food to waste remediation, through natural coastal defences to recreation).

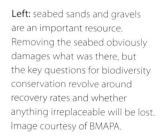

Left: seabed sands and gravels are an important resource. Removing the seabed obviously damages what was there, but the key questions for biodiversity conservation revolve around recovery rates and whether anything irreplaceable will be lost. Image courtesy of BMAPA.

Right: although only a small proportion of the public go scuba diving, many find the marine life fascinating and want to be certain that it is being protected. Photographed in September 2016.

Left: offshore structures involved in extraction or generation for energy disturb localised areas of seabed and create a different substratum in what previously had been only sediment. They can also function as de facto marine protected areas where fishing is prohibited.
Right: seabed species make an important contribution to our diet, but fishing for them needs care to avoid damaging fragile habitats important for biodiversity and to ensure that what is taken is sustainable. Image: Tupungato/Shutterstock

The following section is my brief overview of what is done and what should be done to protect seabed marine life around Britain. I have already described the topic in my book *Marine Biodiversity Conservation: a Practical Approach*. Those professionally involved in conservation will keep up to date with the latest viewpoints and knowledge in a range of journals including *Marine Pollution Bulletin*, *Aquatic Conservation* and *Marine Policy*, amongst many others. Those with a more general interest and concern will be members of relevant voluntary conservation organisations and there will be information in their newsletters and campaigns.

THREATS AND SENSITIVITIES

Threats to what? In the context of this book it is threats to the structure and functioning of seabed ecosystems; to the continued presence of rare, scarce and fragile habitats and species; and to locations, habitats and species that are valued for their intrinsic appeal.

The way that threats are expressed by those professionally involved in marine environmental protection and management may sound like gobbledegook to a lay person. There will be talk of 'pressures', of 'factors', of 'activities' and there will be careful definition of what is meant by the word 'sensitive' and precise use of the words 'fragile' 'threatened' and 'vulnerable'. Fundamentally, an assessment of the likely ecological impact of human activities should ask what habitats and species are in an area being affected by, or likely to be affected by, a particular activity or event; the likely scale of loss; whether those habitats and/or species will recover (and how long recovery will take) if they are adversely affected, and whether those habitats and/or species matter from the point of view of biodiversity conservation.

Below is a list of pressures, with examples of the causative activities that are relevant to subtidal seabeds. They are listed very approximately in the order of likely severity, taking account of the likely persistence of effects. However, the scale and the habitats impacted must be taken into account when assessing the level of severity in a particular case. For instance, the construction of a slipway across a mudflat would not have the scale of impact of frequent disposal of dredge spoil on a subtidal reef system with a high level of biodiversity.

- Coastal developments (e.g. coastal defences/breakwaters; causeways; ports and marinas)
- Physical disturbance (e.g. benthic trawling and dredging, anchoring)
- Eutrophication (e.g. nutrient enrichment via agriculture and sewage disposal)
- Disruption of food webs (e.g. depletion of top predators)
- Introduction of non-native species (e.g. via shipping and aquaculture)
- Pollution by chemicals (e.g. via industrial discharge; land runoff; use of biocides)
- Selective extraction (e.g. via fishing – traps; fishing – recreational)
- Oil pollution (e.g. oil exploration/production platforms, oil spills, oil terminal discharges)
- Sediment extraction (port dredging; sand and gravel extraction)
- Dumping (non-toxic, e.g. from navigational dredging)
- Offshore constructions (e.g. windfarms, pipelines, oil/gas rigs)
- Food production (aquaculture)
- Littering (including plastics)
- Warm water (power stations cooling water effluent)

- Collecting (scientific, educational, curio, souvenir)
- Radioactive discharge (power stations).

There are significant uncertainties about the dangers posed by plastics (especially microplastics and plastic fragments) and about multiple stressors (such as warming, acidification and deoxygenation) acting together.

Degree of threat is assessed by considering the likelihood of a threat occurring and whether the species or habitat being considered will be affected (their vulnerability). The intolerance (lack of resistance) of a species (including that of constituent species of a biotope) to a factor (for instance, physical disturbance, decreased salinity, nutrient enrichment) and the likely recovery (resilience), including rate of recovery after damage has occurred, comes next. Those are the considerations from which sensitivity is assessed. The concepts surrounding assessment of sensitivity were developed and applied at the Marine Biological Association as a part of the Marine Life Information Network (MarLIN) programme there. The resulting information, together with the tools to use it, are available on the MarLIN website. The work continues to be supported by stakeholders including government departments and the statutory nature conservation agencies.

All of the above seems very structured and scientifically based. However, there are difficulties assessing recovery in particular as, for both species and biotopes, we need to ask the question 'recover to what?'. If we are aiming for some pristine state, we do not know what that should be. If we are aiming for the status quo, that may not be anywhere near natural and we should aim higher. Identifying how quickly we should expect recovery to occur is also problematic. For many species, we do not

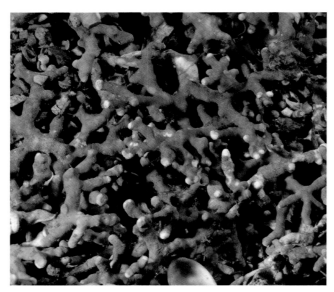

Maerl and maerl beds are highly intolerant (low/no resistance) to damage by 'abrasion and physical disturbance' and their poor recruitment and slow growth means low or no recovery potential. They are therefore classified as 'very high' or 'high' sensitivity to abrasion and physical disturbance. Here, *Phymatolithon calcareum*. Image width *c.* 5cm.

Individuals and reefs of Ross Worm *Sabellaria spinulosa* are intermediate intolerance (low resistance) to damage by 'abrasion and physical disturbance' but rapid recruitment and fast growth means high or medium recovery potential. They are 'low' or 'medium' sensitivity to abrasion and physical disturbance. Image width *c.* 5cm.

Both of the habitats shown above are listed under various directives, conventions and statutes as 'Features' to be protected. In developing management actions due regard needs to be given to their sensitivity to the activities being considered.

Lost or abandoned fishing gear can end-up on the seabed becoming entangled with wildlife. 'Sea fangles' are bundles of fishing net, ropes and organisms such as seaweeds and sea fans that are washed ashore. Here, algae and the skeletons of Pink Sea Fans *Eunicella verrucosa* among monofilament gill net.

have information on their life history traits (reproductive characteristics, how far the adults or dispersal stages can travel before recolonising, their growth rates and their longevity). For many biotopes, it is difficult to identify a few researched species that can be used to represent their sensitivity. Also, only species and biotopes that are designated (see below) may have been researched – there are too many and it takes too long to undertake the research on everything. With all of the gaps in information, it is often knowledge and experience that is brought into play when, for instance, assessing the state of our seas. That approach worked well for the Charting Progress 2 report, which assessed the state of UK seas by bringing together scientists who had knowledge and experience and contributed their wisdom to the exercise. Relevant knowledge can come from lectures and from reading books or scientific papers, but experience requires seeing for yourself seabed wildlife, impacts from human activities and natural change.

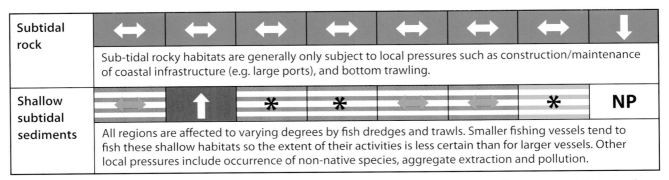

Subtidal rock	↔	↔	↔	↔	↔	↔	↔	↓
	Sub-tidal rocky habitats are generally only subject to local pressures such as construction/maintenance of coastal infrastructure (e.g. large ports), and bottom trawling.							
Shallow subtidal sediments	←	↑	*	*	→	→	*	NP
	All regions are affected to varying degrees by fish dredges and trawls. Smaller fishing vessels tend to fish these shallow habitats so the extent of their activities is less certain than for larger vessels. Other local pressures include occurrence of non-native species, aggregate extraction and pollution.							

The UK's Charting Progress 2 initiative brought together knowledge, information and experienced scientists to provide an assessment of the state of UK seas and to summarise that assessment in an understandable way. Above is a clip showing the assessment for shallow subtidal seabeds. The colours represent 'Few or no problems' (green), 'Some problems' (orange) and 'Many problems' (red). The arrows show trends (up is 'State improving'). Broken colour bands indicate low confidence and the asterisk that no information is available. The columns represent the eight Regional Seas.

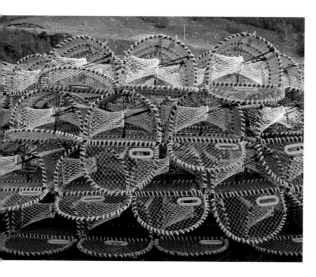

Research has shown that catching lobsters including Norway Lobster (scampi) using pots (creels) usually causes little collateral damage to seabed wildlife. On the other hand, scallop dredges have the potential to destroy seabed habitats and associated species. They are especially damaging where rock reefs occur amongst the sediments being targeted or on some biogenic structures such as maerl beds. Left: scampi creels in Loch Torridon in north-west Scotland. Right: a small scallop dredger in Sutton Harbour, Plymouth.

PROTECTION

In a Utopian world, good stewardship of our seas and their resources would be achieved in relation to all human activities without the need for legislation. In practice, regulation and enforcement is essential. Conservation happens because of public pressure and because governments want to see resources protected. Conservation is achieved through good practice and by the implementation of directives, conventions and statutes that aim to provide protection for habitats and species in our seas. A starting point for creating detailed objectives and targets is having an overarching goal. In the UK, the Government's goal is to have 'Clean, healthy, safe, productive and biologically diverse oceans and seas'.

One of the measures applied to protect seabed habitats and species is the establishment of Marine Protected Areas (MPAs). The EC Habitats Directive is applied across Europe and includes five marine habitats relevant to Britain for designation and protection as Special Areas of Conservation (SAC). Although the Directive (which began being prepared in 1989) pre-dated a well-thought-out biotopes classification for marine habitats and an understanding of what constitutes a threatened marine habitat, it has been interpreted for marine habitats in a way that has protected many sites important for biodiversity around Britain. However, the statutory and other measures aimed at protecting marine biodiversity may not be forever and, with the UK scheduled to leave the European Union, there is concern

Fisheries protection vessels off Lundy in 2007.

about continued protection for seabed habitats established in MPAs as a result of the EC Habitats Directive. Another driver for conservation, the OSPAR Convention for the Protection of the Marine Environment of the North-East Atlantic, calls for 'an ecologically coherent network of well-managed marine protected areas'. That goal has been incorporated into the EU Marine Strategy Framework Directive, but it seems that ecological coherence is almost impossible to define and the concept that networks (with direct connections between separate protected areas) can exist in the marine environment is not supported by scientific studies. Indeed, it is not species with good dispersal capability that need to be protected in MPAs, but those with larvae or propagules that go no distance at all from their parents and need to be looked after where they occur. The sites now protected or scheduled for protection as MPAs in England, Scotland and Wales are a valuable resource for biodiversity conservation but need relevant management measures. Those management measures should be based on the science that we have and the co-operation of all of those with interests in how those areas are used.

The example given in the image below is present in a SAC designated for 'Reefs' and many of the species in it are of high sensitivity to physical disturbance in particular. In order to determine appropriate management measures, the manager needs to know what those species are. However, many of those sensitive species are not listed as being of conservation importance (see the piece on *Bolteniopsis prenanti* on the next page) and therefore are not even candidates to be considered in determining management measures. Protecting this important habitat requires more than just a label, and there need to be effective management measures put into place to protect all of the special and sensitive features in a designated site.

Measures that will ensure the long-term survival of habitats and species range from voluntary agreements to statutory regulations. All need monitoring and effective enforcement.

Some low-lying reef communities may get in the way of mobile bottom-fishing gear and be destroyed. If they include rare and scarce species, or species that are unlikely to return once lost, it matters very much to ensure that they are protected. Here, off the eastern seaboard of the Isles of Scilly, an exceptionally high diversity of fragile species and many species that are nationally rare or scarce including, at the time surveys were undertaken in 2009 and 2010, one new record for Britain and one species that had just been described as new to science. 'Eunicella verrucosa and Pentapora foliacea on wave-exposed circalittoral rock' (A4.1311 / CR.HCR.XFa.ByErSp.Eun). Image width c. 30cm.

There are many plants and animals that do not find their way onto lists of species to be designated for protection but which are rare, scarce, in decline or threatened with decline (sensitive to human activities). The Miniature Sea Onion *Bolteniopsis prenanti* (a sea squirt) is known from only two locations (in the Isles of Scilly) in Britain and is therefore Nationally Rare. It is recorded from only about 10 locations worldwide. It is not on any lists of species to be protected. Safeguarding such species needs policy advisors, site managers and regulators to know of its status and where it occurs and to ensure that at least its habitat is protected from potentially damaging activities. Image width *c.* 4cm.

Some species may benefit from being a designated feature in a protected area, or being subject to protection wherever they occur. The European Spiny Lobster *Palinurus elephas* is a Species of Principal Importance in England and Wales and a Priority Marine Feature in Scotland. It has previously been a Biodiversity Action Plan Priority Species and is a Feature [or Species] of Conservation Importance in the Marine Conservation Zone guidance. This one was photographed in the Isles of Scilly. Image width *c.* 40cm.

MARINE PROTECTED AREAS – WHAT THEY WILL AND WILL NOT DO

MPAs have been established worldwide and for governments, if they are party to relevant directives and conventions, there will be targets to achieve. One such target is the Convention on Biodiversity Aichi Target 11: 'By 2020, at least …. 10% of coastal and marine areas, especially areas of particular importance for biodiversity and ecosystem services, are conserved through effectively and equitably managed, ecologically representative and well-connected systems of protected areas'. At the time of writing, 23% of UK seas were within MPAs of one sort or another. MPAs should be seen in the context of a wider marine environment, 100% of which should be under better management.

There are different types of Marine Protected Areas (MPAs) worldwide summarised in the diagram below.

MAIN TYPES OF MARINE PROTECTED AREAS (MPAs)

MARINE RESERVES (IUCN categories I & II)		FEATURE-BASED MPAs e.g. SAC (EC Habitats Directive), MCZ (UK Marine & Coastal Access Act) (IUCN categories III & IV)	MARINE PARKS (IUCN categories V & VI)
NO USE No human activities permitted.	NO TAKE Strictly protected. Set-aside for protection of biodiversity and for scientific study. Non-extractive and non-damaging activities permitted.	Activities that may damage specific features (habitats and/or species) are prohibited or restricted.	Areas important for their scenic value and where cultural values/activities are protected. Sympathetic use is encouraged. Uses of natural resources are compatible with biodiversity conservation.

MPAs in Britain have many names. Each has its own legislative status, goals and objectives. All can be encompassed within the acronym MPAs. There are Voluntary Marine Conservation Areas (VMCAs), Marine Nature Reserves (MNRs), Special Areas of Conservation (SACs), Marine Conservation Zones (MCZs) (in England and Wales) and Nature Conservation Marine Protected Areas (NCMPAs) (in Scotland). Sites of Special Scientific Interest have been established since the 1950s and a few have been notified for marine biological interest, but in the intertidal zone only.

There is a great danger that MPAs are seen as the only or main measure that will protect valued marine habitats and species. There is an even greater danger that MPAs are designated and that effective management measures are not put into place: they then become known as 'paper parks'. The following text is based on an article that I wrote for the Marine Conservation Society (MCS) magazine, *Marine Conservation*, in autumn 2013.

The record of success for highly protected MPAs enabling the recovery of previously exploited species is outstanding. Lobsters in the No Take Zone off the east coast of Lundy were 7.7 times more abundant than in fished areas just a few years after the No Take Zone was designated. For scallops off the Isle of Man, not only did numbers increase in the protected area, but those individuals will have produced larvae to seed adjacent fished areas and, furthermore, there was a recovery of seabed-attached species where, in the past, heavy mobile fishing gear would have destroyed the community. Recovery may not be immediate – in New Zealand, it took 15 years for the kelp forest to recover in previously over-grazed areas where urchin predators had been fished. The rich communities in muddy gravel off the east coast of Lundy most likely benefit greatly from the lack of scallop dredging since a voluntary agreement in the early 1970s. Importantly, many species that were previously

The seabed around the island of Lundy at the entrance to the Bristol Channel was Britain's first Voluntary Marine Reserve, first statutory Marine Nature Reserve and includes the first No Take Zone for biodiversity conservation in Britain.

Species-rich gravelly sediments that would be degraded by use of heavy mobile fishing gear. Here, at a depth of 30m in Plymouth Sound where such fishing is not permitted, the area is a de facto MPA that is also a SAC. The biotope is most likely: '*Cerianthus lloydii* with *Nemertesia* spp. and other hydroids in circalittoral muddy mixed sediment' (A5.4411 / SS.SMx.CMx. CIIoMx.Nem). Image width *c.* 50cm in the foreground.

MPAs will not safeguard against disease and long-term natural fluctuations in the abundance of species. Here, at Lundy, populations of much-valued soft and hard corals have suffered disease (in the case of at least Pink Sea Fans *Eunicella verrucosa*) and decline in abundance since the mid-1980s. The picture on the left was taken in 1986 and the picture on the right in 2001 in approximately the same location. Image widths *c.* 50cm.

unaffected by whatever was the damaging activity, will remain much as always. Highly protected marine areas (no take, no deposition) are the only way that we are going to be able to track and begin to understand natural fluctuations and the consequences of protecting previously exploited species.

MPAs are one of the tools that we have to protect and restore biodiversity. They have been very effective where previously exploited or damaged seabed species and habitats have become protected. They are particularly needed where rare, scarce or sensitive features are present and where recovery is desired. Highly Protected Marine Areas (marine reserves) further provide the opportunity to describe and better understand natural change where localised human impacts are prevented. MPAs will not protect against widespread water quality changes including increased nutrients and contaminants, against disease, or against non-native species that become pests. They will not prevent changes that may result from seawater warming but, in the shallow seas of the north-east Atlantic, those changes are likely to be a re-balancing of species composition in an area rather than overall loss of biodiversity.

It is all too easy to think that MPAs are only about stopping damaging activities to protect biodiversity. They are also locations where the public can see and enjoy rich marine life, where they can learn about that marine life and where both scientists and natural historians can play a role in understanding how marine ecosystems work.

Diving in the Lundy Marine Nature Reserve in 2005. A fascinating and enjoyable experience enhanced by the availability of guidebooks and advice from wardens.

RESTORATION?

Restoration is frequently employed in terrestrial conservation and involves manipulating habitats and reintroducing species. However, almost all marine habitats and communities are close to natural and do not need manipulation, whilst re-introduction of many seabed species is impractical. As an example, European Spiny Lobsters *Palinurus elephas* were made locally extinct in many areas due to overfishing in the early 1970s and had not recovered since then. In an assessment of possible recovery programmes for marine species made in 2011, actions were suggested to undertake translocation and captive breeding but, in about 2014, widespread recruitment that had never before been recorded took place and populations are now at least partially recovering (see the previous chapter, 'Change'). In the marine environment, recovery is most likely best achieved by removing pressures related to human activities and then left to take place as a natural process.

RESEARCH

Selecting sites for establishment as MPAs – that represent the range of seabed habitats and associated communities of species, and that can protect threatened species and habitats – requires survey data. The seas around Great Britain are rich in survey data (see the earlier chapter, 'Gathering knowledge and creating information – the past 50 years or so') and that information is mostly readily available (see 'Technology takes off'). There are, however, major gaps. Significant survey work is needed to provide the necessary context when determining if a species or a biotope is rare, scarce or threatened, or to determine the location of the biodiversity hotspots, so that representative habitats can be protected. For ecological assessments of the likely sensitivity of any location, survey data will be needed, together with up-to-date touchstones, including lists of rare and scarce species and an understanding of the likely impact of any proposed activities on what is present. There is great danger in feature-led conservation. The lists of designated species in particular are very selective, depending on the availability of quantitative information about occurrence and about abundance. Such quantitative information is often not available and the lists go out of date anyway. Many worthy species that should be taken account of in ecological assessments, and which should be drawn to the attention of managers, are not listed in designated taxa: they could easily be considered as not important. If habitats and species 'of principal importance for the purpose of conserving biodiversity' (Natural Environment and Rural Communities Act of 2006, which refers to habitats and species that need to be taken into account by a public body when performing any of its functions) are to be properly identified, it must be with up-to-date lists determined using relevant criteria, that are not data hungry.

Research is essential, to give context to any concerns about decline or change in habitats and species. That research includes studies of long-term variability, in climate and in species and communities. It includes research to obtain a better understanding of what are viable areas for species and habitats. It includes experimental studies and observations that record their impacts on species or habitats and their recovery from those impacts. It includes studies that reveal life history traits of species and informs assessment of their sensitivity to factors including their potential recovery after a damaging event.

A video still showing the seabed in Herne Bay, Kent on 6 July 2016. The cream-coloured growth is the non-native sea squirt *Didemnum vexillum*. Image: Deborah Phillips. Image width: *c.* 1m.

The most important research is that which helps us to understand what is where, what is rare, what is sensitive and what change is happening, including change brought about by human activity. Bringing together data to create information is of core importance. Some of the ways in which this is being undertaken are outlined in the chapter 'Technology takes off'. All to often, though, we are led by outdated concepts and by outdated or inadequate lists of threatened habitats and species. It is important that science is used effectively and that time and money are spent on actions that will make a difference.

THE DILEMMA OF NON-NATIVE SPECIES

There is a particular dilemma in conservation with regard to non-native species. In 2016, about 90 non-native seabed species were recorded around Britain. Additionally, there will be non-native species that have been here for so long that we consider them native without realising that, perhaps, they came across the Atlantic on the hulls of wooden sailing ships or in ballast in the 17th century. The presence of non-native species degrades naturalness, which is a much-valued feature of a habitat and its associated community of species. Most non-native species fit in, but some become pests, displacing native species and changing habitats. One such is the Carpet Sea Squirt *Didemnum vexillum*, which has the potential to smother native habitats and destroy commercial mussel beds. Some may bring diseases to which native species have no immunity. Can we do anything to exterminate a non-native species from a particular geographical area or can we stop them from spreading? Local extermination may be possible but very expensive to undertake, and may therefore only be done if commercial resources such as mussel beds are threatened. Biosecurity measures can be put into place requiring, for instance, ships coming into an area to have a clean hull. A major vector of importation and spread of non-native species has been mariculture, with many unwanted species being imported accidentally with Pacific Oysters *Magallana gigas* (that have themselves become a pest species in places). Oysters can be placed into quarantine before they are released, but whether new Pacific Oyster farms should be permitted may need to be considered. Most of the measures that can be employed are concerned with preventing or reducing further spread of a species.

WHAT YOU CAN DO

Everyone has a role to play: from those employed as policy advisors through to members of the public passionate about protecting marine biodiversity. That role might be in drafting policy and legislation, in implementing legislation – perhaps through preparing guidance and giving advice – in managing protected sites, in undertaking surveys, in showing support for the protection of wildlife, and in promoting the value of marine biodiversity. So: policy advisors and those in conservation organisations need to understand priorities – what it really matters to protect – and concentrate effort on using scientific knowledge and experience to encourage action. Scientists need to communicate with the public in an understandable and convincing way, setting aside technical language. Those passionate about conservation need to work with the relevant conservation organisations (for instance, The Wildlife Trusts, MCS and the World Wide Fund for Nature (WWF)) to influence decision makers. Those undertaking biological surveys need to ensure that they use techniques that will provide meaningful and accurate information. The results of those surveys should be added to national databases so that they improve the picture that we have of the distribution and character of seabed wildlife. Those who make observations – whether of habitat and species occurrences, of behaviour and life histories of species or of damaging activities – need to submit them to recording schemes. The Marine Biological Association, The Wildlife Trusts and MCS, amongst others, have recording schemes and offer opportunities to participate in citizen science surveys of seabed wildlife. The Seasearch programme, run jointly by MCS and The Wildlife Trusts, uses volunteer scuba divers to fill gaps in knowledge about the sorts of habitats featured in this book.

Contributing to marine conservation action can mean a range of activities, from recording marine life at previously un-surveyed locations to collecting litter. Here, a Seasearch diver records habitats and species in north Cornwall and members of Plymouth Sound BSAC (The British Sub-Aqua Club) collect litter in Plymouth Sound.

Technology takes off

This is very much a 'who would have thought it' chapter, reflecting the great strides that have been made in collecting and disseminating information about what is where: resources that would have astounded marine naturalists just a couple of decades ago. It also looks forward to what we need to carry on doing, and to new developments that will help to fill gaps in our knowledge.

FINDING INFORMATION

The seabed around Britain is very incompletely surveyed for the wildlife that is there. Finding survey data for a particular location or to identify the distribution of a species was once a matter of scouring libraries, but mainly to find survey results in limited circulation reports – journal publishers are rarely interested in publishing the results of yet another survey. When the Marine Nature Conservation Review of Great Britain (MNCR) began in 1987, one of the main objectives of the programme was to access existing survey data and add it to the MNCR database. That was easier said than done! The review of current knowledge in the *MNCR Benthic Marine Ecosystems* volume reflected an early stage in the programme where marine biologists all around Britain were asked what information was available and where it could be found. There were major frustrations with data that were considered commercial and in confidence, and dismay when it transpired that raw survey data had been dumped

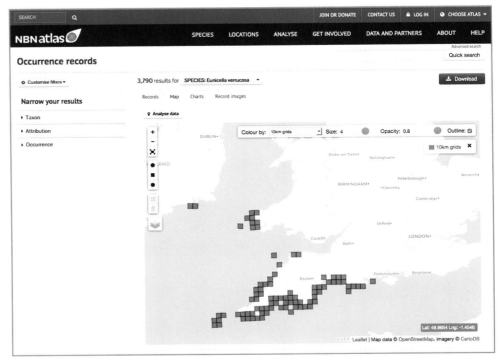

A screenshot from the NBN Atlas showing the recorded distribution of Pink Sea Fans *Eunicella verrucosa* as 10km squares in south-west Britain. Data courtesy of the NBN Atlas with thanks to the organisations that have contributed data: Cornwall Wildlife Trust (Environmental Records Centre for Cornwall and the Isles of Scilly); Dorset Environmental Records Centre; Natural England; Natural Resources Wales; Joint Nature Conservation Committee; Marine Biological Association; Seasearch.

a few years previously. Nevertheless, during the lifetime of the MNCR from 1987 to 1998, a great deal of survey data, both existing and newly gathered, were entered to the database. The MNCR had established a protocol for entering survey data: that protocol and their Advanced Revelation database became a starting point for the development of what is now one of the standard tools for data entry, Marine Recorder. Much else was happening in the late 1990s and, out of discussions that were aimed initially at increasing survey work to gather data, the UK National Biodiversity Network (NBN) came to be. The NBN developed into a scheme, funded mainly by the statutory nature conservation bodies and environmental protection agencies, to draw together existing and new data on the distribution of species, both aquatic and terrestrial, and to develop tools to interrogate and display that data in a way that was useful for recorders and for environmental protection and management. Finding species distributions is now easily achieved through the online NBN Atlas.

Survey work continued in the statutory nature conservation bodies, especially in Scotland and in Wales. Survey work was also being undertaken as a part of ecological impact assessments where development was planned. It was often a requirement of those surveys that data was submitted via Marine Recorder to the NBN and made available for any sort of biotope identification, seabed mapping, etc. Seasearch had taken off and was providing valuable data that helped fill gaps wherever scuba diving could be used. Raw survey data was being translated so that the biotopes present in an area could be identified. The value of what is described as 'collect once, use many times' for survey data was sinking in and is achieved in the UK mainly through the Marine Environmental Data and Information Network (MEDIN), an open partnership committed to improving access to all types of marine data.

Collaborative schemes to ensure common standards, identify validation procedures and develop mapping tools were being initiated, with EU funding, via the European Marine Observation and Data Network (EMODnet) and the Intergovernmental

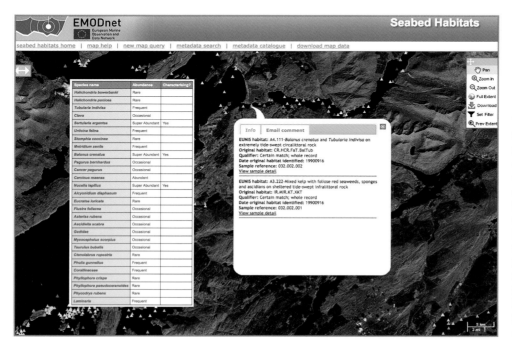

The EMODnet website enables access to survey data including species lists. In this screen shot, a survey site in the Falls of Lora near Oban has been interrogated. The species list on the left was accessed from that page and is pasted onto this image. Information contained here has been derived from data that is made available under the European Marine Observation Data Network (EMODnet) Seabed Habitats project funded by the European Commission's Directorate-General for Maritime Affairs and Fisheries (DG MARE).

Oceanographic Commission (IOC) supported Ocean Biogeographic Information System (OBIS).

So, now, the marine naturalist, nature conservation practitioner, diver or whoever might have an interest in discovering what is where, has two major sources of information: the NBN and EMODnet.

MAKING SEABED BIOLOGY INFORMATION 'USEFUL'

In 1996, I submitted a Viewpoint article to *Marine Pollution Bulletin*: 'Use available data'. The title was from the brief that we always seemed to be given when doing assessments of applications for development. The data were often difficult to access (including being hidden or barred from release) but there was an opportunity to make better use of the data we could find by linking them to assessments of species and biotope sensitivity. By the time the article was published in February 1997, it was dawning on me how such information might be disseminated – a new phenomenon called 'The Internet'.

In the mid-1990s, the then Director of the Marine Biological Association, Michael Whitfield, was developing ideas for electronic dissemination of information and, together, we initiated a programme that was to become the Marine Life Information Network for Britain and Ireland (MarLIN). That all sounds very easy – it was not, and we were greatly helped by a knowledgeable steering group from many interested organisations. The Marine Biological Association was an obvious partner and lead organisation for any marine data initiative. The fact the MarLIN was an early presence on the Internet led to spin-out projects that supported data access and dissemination (the Archive for Marine Species and Habitats Data – DASSH) and the most recent iteration of MarLIN sensitivity information: Marine Evidence-based Sensitivity Assessment (MarESA).

The Marine Life Information Network (MarLIN) provides information for marine environmental management, protection and education. For selected habitats and species, details are given of their ecology, life history traits and likely sensitivity to change, including through human activities.

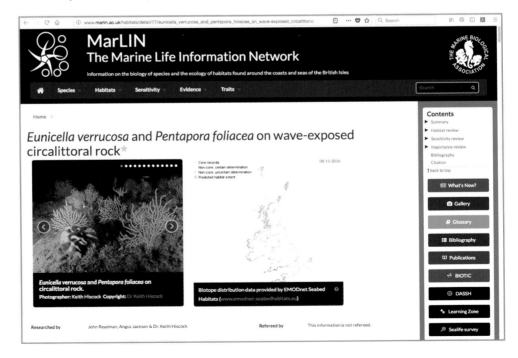

VISUALISATION

This book has been greatly about providing a visual experience, through photographs and line drawings, of the hidden seabed world. Opportunities to make armchair underwater tourists of us abound. The reports produced by Seasearch of their regional surveys are illustrated with high quality images. The statutory nature conservation agencies use images extensively on their websites and brochures. There are beautifully illustrated books by the likes of Paul Naylor, Paul Kay and Frances Dipper as well as the Seasearch series of marine life identification guides. In the future, we should be able to go to maps or charts on a computer screen and to click on a location to see what the seabed looks like there – whether just the topography or, where underwater visibility has allowed, the seabed communities. The DORset Integrated Seabed Study, run by the Dorset Wildlife Trust in 2008 and 2009, undertook such a visualisation project that enables the user to see, on a multibeam sonar backdrop, what the seabed looks like at specific locations.

Much is now being done to take the non-diver to the seabed, to see it all in moving pictures. Resources like YouTube are alive with videos – some much better than others. Television channels often try to include underwater themes and footage in documentaries, natural history programmes and in more lightweight popular broadcasts – some do it much better than others. However, it is important to remember that many of the seas around Britain are highly turbid, and getting

A screen clip from the DORset Integrated Seabed Study offering the user the opportunity to see images from specific locations off Portland Bill. These images can be enlarged by clicking on them. Seabed imagery © Dorset Wildlife Trust. Contains MCA data © Crown copyright.

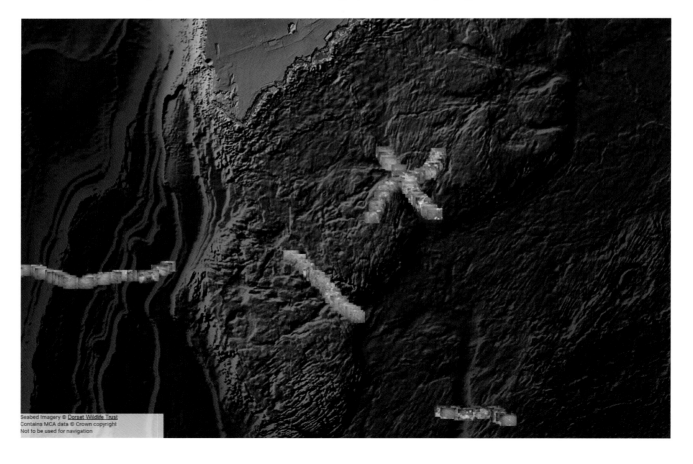

Seabed Imagery © Dorset Wildlife Trust
Contains MCA data © Crown copyright
Not to be used for navigation

Good quality underwater videos reveal the true nature of the seabed around Britain. Here, David Ainsley on a dive in the Firth of Lorn in May 2016.

A topographical survey by multi-beam sonar of the part of the Menai Strait where tidal currents are strongest. The view is towards the east. Blue areas are intertidal. The colours represent depth and are from red (shallow) to green (deep – about maximum 7m below chart datum in this image). The main channel is clearly shown with areas of steeply sloping bedrock. Image courtesy of Mike Roberts (SEACAMS R&D Project Manager at the Marine Centre, Wales, Bangor University).

good wide-angle images may be challenging. For the armchair underwater tourist, impressive software now enables the outputs from multibeam sonar to be manipulated into a virtual fly-through of our underwater reefs, cliffs, canyons and other features. There is no biology in them, but such 3D images give the viewer a sense of how varied and exciting underwater landscapes can be. Draping video images on high resolution terrain models from such swathe bathymetry can now give the viewer a picture of at least broadscale biology.

Computer-generated imagery may give the viewer an accurate idea of the appearance of seabed species and communities. Here, a CGI view of dense Fan Mussels *Atrina fragilis* in a screen clip based on drop-down images collected by Marine Scotland in the Sound of Canna. Reproduced with the permission of Scottish Natural Heritage. Image: Scottish Natural Heritage.

COLLECTING INFORMATION

The gaps in data availability for seabed marine biology at particular locations – even in the shallow seas around Britain – are enormous. Those gaps are not being filled by predictive algorithms, which give rise to attractive coloured maps that nearly always prove to be wrong. Neither are they filled by acoustic surveys (such as acoustic ground discrimination systems or interpreting backscatter from multi-beam surveys) which, again, are very poor predictors of biology (although the ground truthing is often very meaningful). In situ surveys are still needed. They can be undertaken by remote sampling (grabs and dredges), the use of towed photographic/video sledge, remote operated vehicles (ROVs), drop-down video cameras, and by diving. Furthermore,

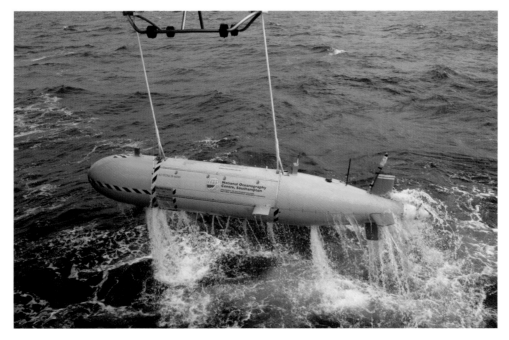

Autosub6000 being recovered from the water. The image was taken during the RRS *James Cook* cruise off south-west Britain in autumn 2015, which included work at Haig Fras. Image courtesy of Russell Wynn, Chief Scientist at the National Oceanography Centre's Marine Autonomous and Robotic Systems (MARS) Innovation Centre.

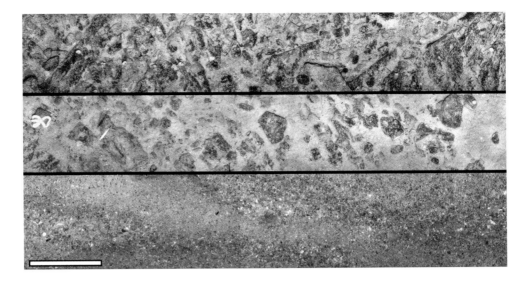

Tiled images from *Autosub6000* taken in the region of Haig Fras, off west Cornwall, and illustrating different seabed types. From work undertaken by Noelie Benoist and others at the National Oceanography Centre, Southampton. The habitats are described as: 'Hard'; 'Hard + sand', and 'Coarse'. The scale bar represents 50cm.

new and exciting tools are becoming available. They include, in particular, Autonomous Underwater Vehicles (AUVs) that may one day patrol the level seabed photographing what is there, wherever underwater visibility and the strength of waves and tidal streams allows. The image shown above is taken from the *Autosub6000*, described (by Brian Bett) as 'an excellent tool for broadscale assessment of the seafloor environment and associated fauna in areas of modest topographic variation'. He goes on to observe, 'What we would call hover class vehicles are needed to tackle features such as Haig Fras, and sublittoral rock generally. Such machines do exist and are certain to revolutionise research and monitoring in tricky habitats'.

Whilst remote devices get more and more sophisticated and widely used, it is direct observations by divers and those viewing high quality images from manoeuvrable devices that will continue to reveal with absolute certainty what is where at particular locations. Divers will see species not previously recorded from Britain, will notice losses and gains in abundance of species and habitats and will witness behaviours such as breeding and settlement. Every record of these observations enhances our knowledge by adding another piece to the jigsaw.

Glossary

The glossary includes technical terms and abbreviations or acronyms used more than once in the text. Terms used once are explained in the text.

Aggregates (in relation to seabed use) Sediment extracted from the seabed including sand, gravel, pebbles. Cf. Sand, gravel, pebble

Algae Unicellular or multicellular organisms in the Kingdom Plantae that have chlorophyll and other pigments but lack true stems, roots, and leaves. On the seabed, they are conspicuously red, green and brown seaweeds.

Algorithm A set of rules to be followed in calculations or other problem-solving operations, especially by a computer. Predictive algorithms based on such characteristics as wave exposure and depth have been used to identify seabed habitats but the results have very low confidence.

Alien species A non-established introduced species, which would be incapable of establishing self-sustaining or self-propagating populations in the new area without human interference. Cf. 'Introduced species', 'Non-native'

Amphipod Belonging to the Phylum Arthropoda, Subphylum Crustacea and Order Amphipoda: a group of crustaceans recognised by their laterally compressed bodies, lack of a carapace, and numerous, differently modified legs.

Artificial reef A structure placed on the seabed with the intention of increasing biodiversity, but often as an addition to other purposes. For instance: coastal defences; oil rigs dropped to the seabed as a cheaper alternative to retrieval; concrete structures that incorporate human ashes (bereavement balls), and vessels sunk to provide recreational opportunities for divers.

Ascidian A sea squirt belonging to the Phylum Chordata and Class Ascidiacea. This marine animal which lives attached to rocks. The larvae have a primitive backbone called a notochord and sea squirts are considered to be closely related to vertebrates.

Assemblage A generic term used chiefly by some British marine ecologists which does not assume interdependence within a community or association, but appears to have the same broad definition as 'community'. Cf. Community

Barnacle Belonging to the Phylum Arthropoda and Subphylum Crustacea. Two forms are typical. Goose barnacles hang from the substratum by a leathery stalk with the rest of the body protected (to varying degrees) by calcareous shell plates. Acorn barnacles are attached directly to the substratum and protected by tightly fitting calcareous shell plates.

Bedrock Any stable hard substratum, not separated into boulders or smaller sediment units. Cf. Boulder

Benthic On or within the seabed.

Benthos / benthic species Those organisms attached to, or living on, in, or near the seabed.

Biodiversity The number and variety of plants and animals.

Biogenic reef Produced by living organisms.

Biogeography The branch of biology concerned with the geographical distribution of plants and animals, and the factors affecting their distribution.

Biota The animal and plant life of a particular region or habitat. Cf. Epibiota

Biotope The physical 'habitat' with its biological 'community'; a term which refers to the combination of physical environment (habitat) and its distinctive assemblage of conspicuous species.

Bivalve Belonging to the Phylum Mollusca and Class Bivalvia. A type of mollusc that has a shell of two halves, such as clams, mussels and scallops.

Boreal Cool or cold temperate regions of the northern hemisphere. The centre of the boreal region lies in the North Sea. It is bounded by the Boreal–Arctic region from Shetland northwards and in the north-west and south-west of Britain by the Boreal–Lusitanian region. Cf. Biogeography

Boring Makes an excavation (through physical or chemical action) in which to live (for instance, a piddock boring into soft rock) or to access food (for instance, a whelk boring into a mussel).

Boulder An unattached rock, defined in three categories: very large (>1,024mm); large (512–1,024mm); small (256–512mm).

Brachiopod Belonging to the Phylum Brachiopoda. A marine animal that secretes a shell consisting of two parts (valves) that enclose an arm-like lophophore used for feeding.

Brackish Referring to mixtures of fresh and seawater. Usually regarded as being a salinity of between 0.5 and 30. Cf. Salinity

Brittlestar Belonging to the Phylum Echinodermata and Class Ophiuroidea. Brittlestars are related to starfish but recognised by their long, thin and articulated arms, which (as the name suggests) break very easily.

Bryozoan / Bryozoa Belonging to the Phylum Bryozoa. A marine animal that is made up of a tiny colony of individuals that grow attached to rocks or on seaweed. Examples are sea mats and hornwracks. Cf. Sea mat

Cave Strictly, a hollow normally eroded into a cliff, with the penetration being greater than the width at the entrance.

Caves can also be formed by boulders. The UK interpretation of 'caves' for implementation of the EC Habitats Directive considers them to take the form of tunnels or caverns, with one or more entrances, in which vertical and overhanging rock surfaces form the principal marine habitat.

Chalk A soft, fine-grained sedimentary rock, normally white, consisting almost entirely of calcium carbonate.

Characteristic 1) A distinguishing feature, the distinctive state or expression of a character. 2) Of species: special to or especially abundant in a particular situation or biotope. Characteristic species should be immediately conspicuous and easily identified.

Chart datum A set reference point on charts for water depth in relation to tides. On metric charts for which the UK Hydrographic Office is the charting authority, chart datum is a level as close as possible to Lowest Astronomical Tide.

Chiton A mollusc belonging to the Phylum Mollusca and Class Polyplacophora. It lives attached to rocks. Its shell is made up of several plates like a suit of armour. Chitons are also called coat-of-mail shells.

Circalittoral The animal-dominated subzone of the rocky sublittoral below that dominated by algae. It is divided into 1) the upper circalittoral: the part of the circalittoral subzone on hard substrata distinguished by the presence of scattered foliose algae amongst the dominating animals with its lower limit being the maximum limit of depth for foliose algae, and 2) the lower circalittoral: the part of the circalittoral subzone on hard substrata below the maximum depth limit of foliose algae. Cf. Infralittoral

Climate The totality of the weather conditions at a certain location over a certain period (conventionally 30 years).

Cobble A rock fragment that is smaller than a boulder and larger than a pebble. Defined by two categories: large (128–256mm); small (64–128mm).

Commensal An association with another species benefitting one species whilst the other species is not adversely affected. Cf. Parasite

Community A group of organisms occurring in a particular environment, presumably interacting with each other and with the environment, and identifiable from other groups by means of ecological survey. Cf. Assemblage

Conservation (of nature) The regulation of human use of the global ecosystem to sustain its diversity of content indefinitely, including restoration where necessary.

Coralline Relating to, or resembling, coral, especially any calcareous red alga impregnated with calcium carbonate.

Crevice A narrow crack in a hard substratum <10mm wide at its entrance, with the penetration being greater than the width at the entrance. Crevices often support a distinct community of species. Cf. Fissure

Crustacean Belonging to the Subphylum Crustacea. A group of marine invertebrates that have a rigid exoskeleton (skin) and jointed legs, including crabs, lobsters, barnacles and prawns.

Current The flow of water in a particular direction. In oceanography, 'current' refers to residual flow after any tidal element (i.e. tidal streams/tidal currents) has been removed.

Diversity The state or quality of being different or varied. Species diversity can be expressed by diversity indices, most of which take account of both the number of species and number of individuals per species. In conservation assessment: the richness of different types in a location, including the number of different biotopes and numbers of species. The number of species present in an example of a particular biotope.

Dredge Bottom sampling equipment towed along the seabed for collecting benthic sediment and organisms. Dredges are also used for the commercial collection of benthic organisms, e.g. scallops.

Echinoderm Belonging to the Phylum Echinodermata. Echinoderms usually have a symmetry of five and include starfish, sea urchins, sea cucumbers and featherstars.

Ecology The study of how animals, plants and all other organisms interact with the environment.

Ecosystem A community of organisms and their physical environment interacting as an ecological unit. Usage can include reference to large units such as the North Sea down to much smaller units such as kelp holdfasts.

Enclosed coast A marine inlet or harbour fully enclosed from the open sea except at the entrance, not normally open to the sea at two ends. The connection with the open sea is normally less restricted than is the case with lagoons. Cf. Lagoon

Endemic Native to, and restricted to, a particular geographical region or location.

Environment The surroundings or conditions in which an organism lives and that affect it. Cf. Habitat

Ephemeral Lasting for a short time.

Epibenthic Living on the surface of the seabed. Cf. Epifauna, Epiflora

Epibiotic/Epibiota Living attached to the surface of another organism, without any detriment or benefit to the host. Cf. Biota

Epifauna Animals living on the surface of the substratum. Cf. Infauna, Epibenthic

Epiflora Plants living on the surface of the substratum. Cf. Epibenthic

Estuary A semi-enclosed coastal body of water which has a free connection with the open sea, and within which sea water is measurably diluted by fresh water derived from land drainage.

Exposed Of wave action. Prevailing wind is onshore although there is a degree of shelter because of extensive shallow areas offshore, offshore obstructions, a restricted (<90°) window to open water. These sites will not generally be exposed to strong or regular swell. This can also include open coasts facing away from prevailing winds but where strong winds with a long fetch are frequent. Cf. Extremely exposed, Very exposed, Moderately exposed, Sheltered, Very sheltered, Extremely sheltered, Ultra-sheltered

Extremely exposed Of wave action. Open coastlines which face into prevailing wind and receive oceanic swell without any offshore breaks (such as islands or shallows) for several thousand kilometres and where deep water is close to the shore (50m depth contour within about 300m). Cf. Very exposed, Exposed, Moderately exposed, Sheltered, Very sheltered, Extremely sheltered, Ultra-sheltered

Extremely sheltered Of wave action. Fully enclosed coasts with a fetch no greater than about 3km. Cf. Extremely exposed, Very exposed, Exposed, Moderately exposed, Sheltered, Very sheltered, Ultra-sheltered

Fauna 1) The animal life of a given region. 2) A descriptive catalogue of 1). Cf. Flora

Favourable conservation status Achieving 'Favourable conservation status' is a core objective of the EC Habitats Directive. Interpreted as habitats having sufficient area and quality and species have a sufficient population size to ensure their survival into the medium to long term, along with favourable future prospects in the face of pressures and threats.

Feature (conservation) Used to refer to specific habitats and species that may characterise a location and to habitats and species that are listed as of marine natural heritage importance.

Fetch The distance of open sea over which the wind blows to generate waves. Cf. Exposed, Extremely exposed, Very exposed, Moderately exposed, Sheltered, Very sheltered, Extremely sheltered, Ultra-sheltered, Wave exposure

Fissure A crack in a hard substratum >10mm wide at its entrance, with the depth being greater than the width at the entrance. Cf. Crevice, Cave, Gulley

Flora 1) The plant life of a given region. 2) A descriptive catalogue of 1). Cf. Fauna

Genus A taxonomic category that encompasses a group of very closely related species of plants or animals ranking below 'Family' and above 'species'.

Geomorphology The branch of geology concerned with the structure, origin and development of topographical features of the earth's crust. Cf. Physiographic

Grab A mechanical bottom-sampling device which is lowered vertically from a stationary ship, for collection of sublittoral sediment and infauna. Cf. Dredge, Trawl

Gravel Sediment particles 4–16mm in diameter which may be formed from rock, shell fragments or maerl. Cf. Sand

Ground truth Information provided by direct observation as opposed to information provided by inference.

Gully A vertical space between two rock walls at least 0.5m wide and 0.5m or more in depth. Cf. Fissure, Cave

Habitat The place in which a plant or animal lives. It is defined for the marine environment according to geographical location, physiographic features and the physical and chemical environment (including salinity, wave exposure, strength of tidal streams, geology, biological zone, substratum, features (e.g. crevices, overhangs, rockpools) and modifiers (e.g. sand-scour, wave-surge, substratum mobility). Cf. Environment

Holdfast An attachment structure that anchors macroalgae to the substratum, which resembles a collection of roots but has no nutrient-gathering role. Cf. Macroalgae

Holothuroidea / Holothurian Belonging to the Phylum Echinodermata and Class Holothuroidea. Elongate, cylindrical animals, tapered at each end with a mouth surrounded by retractile feeding tentacles: sea cucumbers.

Hydrography The scientific study of seas, lakes and rivers.

Hydrozoa / Hydroid Belonging to the Phylum Cnidaria and Class Hydrozoa. A type of marine invertebrate (cnidarian) related to anemones, corals and jellyfish. They can often look like plants but they are in fact animals. Includes 'sea firs' and 'white weeds'. Cf. Sea fir

Infauna Benthic animals which live within the seabed. Cf. Epifauna

Infralittoral A subzone of the sublittoral in which upward-facing rocks are dominated by erect algae. It can be further subdivided into the upper infralittoral (dominated by kelps) and lower infralittoral (dominated by foliose algae). In heavily grazed situations, the rock may be dominated by encrusting algae.

Inlet See 'Marine inlet'.

Introduced species Any species which has been introduced directly or indirectly by human agency (deliberate or otherwise), to an area where it has not occurred in historical times and which is separate from and lies outside the area where natural range extension could be expected (i.e. outside its natural geographical range).

Kelp A group of large, brown algae of the Order Laminariales, common in the sublittoral fringe and infralittoral zone.

Knot (of velocity) A unit of speed equal to one nautical mile (1.852km) per hour, approximately 1.151mph.

Lagoon (saline) A shallow body of coastal salt water (from brackish to hypersaline) partially separated from an adjacent sea by a barrier of sand or other sediment, or less frequently, by rocks. Cf. Enclosed coast

Larva A juvenile phase differing markedly in morphology and ecology from the adult.

Lusitanian Referring to a biogeographical region centred to the south of the British Isles and influencing the extreme south-west of the British Isles. Cf. Boreal

Macroalgae Multicellular macroscopic benthic algae, commonly known as seaweeds.

Macrofauna The larger organisms of the benthos, exceeding 1mm in length. Cf. Meiofauna

Maerl Twig-like unattached (free-living) calcareous red algae, often a mixture of species and including species which form a spiky cover on loose, small stones – 'hedgehog stones'.

Mariculture The cultivation of marine organisms in seawater by human effort for commercial purposes.

Marine inlet An indentation of the shoreline including estuaries, enclosed bays and the sounds, straits and narrows between land masses.

Marine Relating to the sea.

Meiofauna Small animals which pass through a 1mm-mesh sieve but are retained by a 0.1mm mesh. Cf macrofauna

Migration Movement from one region or habitat to another according to the seasons.

Moderately exposed Of wave action. Open coasts facing away from prevailing winds and without a long fetch but where strong winds can be frequent. Cf. Extremely exposed, Very exposed, Exposed, Sheltered, Very sheltered, Extremely sheltered, Ultra-sheltered

Moderately strong Of tidal currents. Surface flow is up to 1–3 knots. Cf. Very strong, Moderately strong, Weak, Very weak

Mollusc Belonging to the Phylum Mollusca. A group of marine invertebrates that are not segmented and usually have a shell, such as cockles, mussels and whelks, though some, like cuttlefish, squid and octopuses, do not.

Monitoring Observing, testing and supervising something periodically to see if there have been any changes. For example, the way in which we study the habitats and animals of the seabed to see if their number and distribution changes, so we can look at what we think is causing those changes.

MPA / Marine Protected Area Any area of intertidal or subtidal terrain, including geological and geomorphological features, together with its overlying water and associated flora, fauna, historical and cultural features, which has been dedicated and managed, through legal or other effective means, to achieve the long-term protection of nature with associated ecosystem services and cultural values.

Mud Fine particles of silt and/or clay, <0.0625mm diameter. Cf. Sand, Silt

Mysid Belonging to the Phylum Arthropoda, Subphylum Crustacea. Slender, shrimp-like crustaceans with translucent bodies, feathery appendages, a broad tail fan and obvious eyes. Known as opossum shrimps.

Natural history The systematic study of natural organisms through observation – what plants and animals do, how they react to each other and their environment, how they are organised into larger groupings like populations and communities. Those who study natural history are known as 'naturalists' or 'natural historians'.

Naturalness In conservation assessment – the extent to which a location and its associated biotopes is unaffected by anthropogenic activities.

Nautical mile A unit of distance used in navigation, equivalent to 1° of latitude. The standard, or international, nautical mile is 1,852 metres.

Neap tide The astronomical tide of minimum range, occurring at the time of the first and third quarters of the moon.

Non-native (species) A species which has been introduced directly or indirectly by human agency (deliberate or otherwise), to an area where it has not occurred in recent times (about 5,000 years BP) and which is separate from and lies outside the area where natural range extension could be expected.

Open coast Any part of the coast not within a marine inlet, strait or lagoon, including offshore rocks and small islands.

Parasite An organism that lives in or on another living organism (the host), from which it obtains food and other requirements. The host does not benefit from the association and is usually harmed by it. Cf. Commensal

Pebble Rock particle 16–64mm in diameter.

Phylum A group of closely related plants or animals, such as molluscs or echinoderms.

Physiographic Features and attributes of the earth's surface. Cf. Geomorphology

Plankton Tiny organisms which float or swim in the sea: they may be plant-like (phytoplankton) or animals (zooplankton).

Pollution The introduction by man, directly or indirectly, of substances or energy into the marine environment (including estuaries) resulting in such deleterious effects as harm to living resources, hazards to human health, hindrance to marine activities including fishing, impairment of quality for use of seawater and reduction in amenities.

Polychaeta / polychaetes Belonging to the Phylum Annelida and the Class Polychaeta. Segmented worms, characterised by extensions of each segment called 'parapodia' that bear bundles of bristles, hence the term 'many bristled' or 'poly' 'chaeta'.

Polyp (of zoology) The hollow, columnar, soft tissue of sea anemones, corals and sea firs that has tentacles.

Population All individuals of one species occupying a defined area and usually isolated to some degree from other similar groups.

Porifera Belonging to the Phylum Porifera. Sponges.

Predator An organism that feeds by preying on other organisms, killing them for food.

Pressure (in relation to environmental protection) A term that is used to describe an event or human activity that may lead to a risk or hazard to marine ecosystem integrity including to aspects of biodiversity.

Propagule Part of an plant that can give rise to a new organism, especially algal spores.

Pycnogonida Belonging to the Phylum Chelicerata and the Class Pycnogonida. Sea spiders have slim bodies divided into a head and a segmented trunk with eight, often slender, legs that give them a superficial resemblance to true spiders.

Rapids Strong tidal streams resulting from a constriction in the coastline at the entrance to, or within the length of, an enclosed body of water such as a sea loch. Depth is usually shallower than 5m.

Recovery A passive event that relies on recolonisation and regrowth of pre-existing species and an associated rebalancing of ecosystem structure and functioning.

Regime shift (in communities) A shift in community structure and composition at a location or in regional seas.

Representativeness In conservation assessment, typical of a feature, habitat or assemblage of species.

Ria A drowned river valley in an area of high relief. Most have resulted from the post-glacial rise in relative sea level and are predominantly marine throughout.

SAC Special Area of Conservation.

Salinity A measure of the concentration of dissolved salts in seawater. Until recently, salinity was expressed as parts per thousand (ppt or ‰) but is now referred to as 'practical salinity units' (psu). In most cases, where a high degree of accuracy is not required, old and new figures for salinity can be used interchangeably. Freshwater is regarded as <0.5 (limnetic), seawater as >30 (euhaline), and brackish water as intermediate.

Sand Particles defined in three size categories: very coarse sand and granules (1–4mm); medium and coarse sand (0.25–1mm); very fine and fine sand (0.062–0.25mm). Cf. Gravel, Mud

Scavenger A type of animal which eats animal or plant remains that it did not kill itself. Cf. Predator

Sea fir A branching hydrozoan. Cf Hydrozoa / Hydroid

Sea loch In Scotland – a marine inlet entered by the tide (on each cycle), and with a salinity generally greater than 30. Brackish conditions may be periodically established, particularly in the surface layers.

Sea mat An encrusting bryozoan. Cf. Bryozoan / Bryozoa

Seagrass A flowering plant (Phylum Angiospermophyta) that is adapted to living fully submerged and rooted in estuarine and marine environments.

Sedentary Attached to a substratum but capable of movement across or through it. Cf. Sessile

Sediment Particulate solid material accumulated by natural processes such as gravel, sand and mud.

Sensitivity /Sensitive An assessment of the intolerance of a species or habitat to damage from an external factor and the time taken for its subsequent recovery.

Sessile Permanently attached to a substratum. Cf. Sedentary

Sheltered Of wave action. Coasts with a restricted fetch and/or open water window. Coasts can face prevailing winds but with a short fetch (say <20km) or extensive shallow areas offshore or may face away from prevailing winds. Cf. Extremely exposed, Very exposed, Exposed, Moderately exposed, Very sheltered, Extremely sheltered, Ultra-sheltered

Sill Lowest point on an underwater ridge or saddle at a relatively shallow depth, separating a basin from an adjacent sea or another basin In sea lochs, sills are structures commonly formed by glaciation, found at the mouth or elsewhere along the length of the loch. Such a threshold can limit water exchange.

Silt Fine-grained sediment particles ranging in size from 0.004mm to 0.0625mm. Cf. Mud

Slack water Period in a tidal cycle when the strength of tidal streams is near zero.

Sonar The use of sound pulses to detect objects or oceanographic features underwater. The sound pulses are reflected and are recorded on their return to the surface.

Sound (physiographic feature) A wide expanse of water.

Species Different plants and animals are species. A species is a group of organisms that all look the same and together are capable of reproducing, producing viable offspring (young).

Spring tide The astronomical tide of maximum range, occurring at or just after new moon and full moon.

Stipe The stalk of an alga that supports the frond(s).

Strait Any deep (>5m) tidal channel between two bodies of open coastal water. Strictly, a strait is the stretch of water between an island and its mainland (or adjacent islands).

Strong (of tidal currents) Surface flow is up to 3–6 knots. Cf. Very strong, Moderately strong, Weak, Very weak

Sublittoral The zone exposed to air at its upper limit only by the lowest spring tides. The sublittoral extends, in temperate zones, from the upper limit of the large kelps and includes, for practical purposes in nearshore areas, all depths below the littoral zone. Cf. Subtidal, Infralittoral, Circalittoral

Substratum Material available for colonisation by plants and animals. A more correct term in this context than 'substrate'.

Subtidal A physical term for the seabed below the mark of Lowest Astronomical Tide. Cf. Sublittoral

Surge gully A narrow marine inlet on a small scale, usually formed by erosion of a rocky shoreline on exposed coasts. Its aspect, facing into waves, and its funnel effect, means that waves entering a surge gulley become higher and of shorter wavelength, causing back-and-forth or multi-directional water movements of considerable force.

Survey A single event at a particular location with the objective of describing the character of that area, site or feature.

Suspension feeders Any organisms which feed on particulate organic matter, including plankton, suspended in the water column.

Swell Sea waves that have left the area where they were generated by the wind, or that have remained after the generating wind has disappeared.

Thermocline A horizontal boundary layer in the water column in which temperature changes sharply with depth.

Tidal current The alternating horizontal movement of water associated with the rise and fall of the tide.

Trawl Equipment towed behind a vessel over the seabed for commercial fishing or scientific collecting.

Ultra-sheltered (of wave exposure) Fully enclosed coasts with a fetch measured in tens or at most a few hundred metres. Cf. Extremely exposed, Very exposed, Exposed, Moderately exposed, Sheltered, Very sheltered, Extremely sheltered

Very exposed (of wave exposure) 1) Open coasts which face into prevailing winds and which receive wind-driven waves and oceanic swell without any offshore obstructions for several hundred kilometres, but where deep water is not close to the shore (50m depth contour further than about 300m). 2) Open coasts adjacent to extremely exposed sites but which face away from prevailing winds. Cf. Extremely exposed, Exposed, Moderately exposed, Sheltered, Very sheltered, Extremely sheltered, Ultra-sheltered

Very sheltered (of wave exposure) Coasts with a fetch of less than about 3km where they face prevailing winds or about 20km where they face away from prevailing winds, or which have offshore obstructions such as reefs or a narrow (<30m) open water window. Cf. Extremely exposed, Very exposed, Exposed, Moderately exposed, Sheltered, Extremely sheltered, Ultra-sheltered

Very strong (of tidal currents) Surface flow can be more than 6 knots. Cf. Strong. Moderately strong, Weak, Very weak

Very weak (of tidal currents) Surface flow is negligible. Cf. Very strong, Strong. Moderately strong, Weak

Voe A ria (in Shetland). A long, narrow inlet that most likely started as a small river valley but that was further deepened by glaciers and eventually flooded after the last glaciation. Cf. Ria

Wave exposure The degree of wave action on an open coast, governed by the fetch and the strength and incidence of the winds. Cf. Fetch, Extremely exposed, Very exposed, Exposed, Moderately exposed, Sheltered, Extremely sheltered, Ultra-sheltered

Weak (of tidal currents) Surface flow is less than 1 knot. Cf. Very strong, Strong, Moderately strong, Very weak

Biotopes listed and illustrated

The following biotopes are in alphabetical order of their name. The EUNIS (European Nature Information System) and JNCC (Joint Nature Conservation Committee) codes are given. The names and codes were up to date in December 2017. How biotopes are determined and coded is described on pages 36–39.

Biotope	Page
'*Abra alba* and *Nucula nitidosa* in circalittoral muddy sand or slightly mixed sediment' (A5.261 / SS.SSA.CMuSa.AalbNuc)	68
'*Alaria esculenta* on exposed sublittoral fringe bedrock' (A3.111 / IR.HIR.KFaR.Ala)	78
'*Alaria esculenta* forest with dense anemones and crustose sponges on extremely exposed infralittoral bedrock' (A3.112 / IR.HIR.KFaR.AlaAnCrSp)	43
'*Alcyonium digitatum* and *Metridium dianthus* on moderately wave-exposed circalittoral steel wrecks' (A4.721 / CR.FCR.FouFa.AdigMdia)	114
'*Amphiura filiformis*, *Kurtiella bidentata* and *Abra nitida* in circalittoral sandy mud' (A5.351 / SS.SMu.CSaMu.AfilMysAnit)	71, 73
'Angiosperm communities in reduced salinity' (A5.54 / SS.SMp.Ang)	184
'*Antedon* spp., solitary ascidians and fine hydroids on sheltered circalittoral rock' (A4.313 / CR.LCR.BrAs.AntAsH)	137
'*Aphelochaeta marioni* and *Tubificoides* spp. in variable salinity infralittoral mud' (A5.322 / SS.SMu.SMuVS.AphTubi)	171, 172
'*Arenicola marina* in infralittoral fine sand or muddy sand' (A5.243 / SS.SSa.IMuSa.AreISa)	62, 72
'Atlantic and Mediterranean moderate energy infralittoral rock' (A3.2 / IR.MIR). (Level 4 and 5 biotopes are listed separately)	105
'*Balanus crenatus* and *Tubularia indivisa* on extremely tide-swept circalittoral rock' (A4.111 / CR.HCR.FaT.BalTub)	162, 166, 168
'*Beggiatoa* spp. on anoxic sublittoral mud' (A5.7211 / SS.SMu.IFiMu.Beg)	161
'Bryozoan turf and erect sponges on tide-swept circalittoral rock' (A4.131 / CR.HCR.XFa.ByErSp). (Level 6 biotopes are listed separately)	38, 101, 102, 104, 235
'Burrowing megafauna and *Maxmuelleria lankesteri* in circalittoral mud' (A5.362 / SS.SMu.CFiMu.MegMax)	76
'*Caryophyllia* (*Caryophyllia*) *smithii* and *Swiftia pallida* on circalittoral rock' (A4.211 / CR.MCR.EcCr.CarSwi)	81
'*Cerianthus lloydii* with *Nemertesia* spp. and other hydroids in circalittoral muddy mixed sediment' (A5.4411 / SS.SMx.CMx.ClloMx.Nem)	63, 238
'Circalittoral coarse sediment' (A5.14 / SS.SCS.CCS). (Level 5 biotopes are listed separately)	64
'Circalittoral faunal communities in variable salinity' (A4.25 / CR.MCR.CFaVS). (Level 4 and 5 biotopes are listed separately)	174
'Circalittoral mixed sediments' (A5.44 / SS.SMx.CMx). (Level 4 and 5 biotopes are listed separately)	161
'Communities on soft circalittoral rock' (A4.23 / CR.MCR.SfR)	85
'*Cordylophora caspia* and *Einhornia crustulenta* on reduced salinity infralittoral rock' (A3.362 / IR.LIR.IFaVS.CcasEin)	179

Index

A

B

Acknowledgements

I have mentioned, wherever possible in the text, those scientists, divers and various colleagues and friends who have helped me to explore and try to understand seabed marine life over the past 50 years. There are organisations, too, that have been of key importance in educating me, employing me or commissioning research. It all started at Westfield College (University of London) where I was fortunate, as an undergraduate, to be helped in pursuing diving research by Minnie Courtney, a Senior Lecturer and fellow diver. I have already acknowledged the patience of staff and colleagues at the Marine Science Laboratory at Menai Bridge (now part of the University of Bangor) with my determined attitude, which led to some breaking of rules and misuse of equipment during the course of my PhD studies and subsequently while I hung on there. I was then fortunate to be employed by the Field Studies Council Oil Pollution Research Unit at Orielton Field Centre in Pembrokeshire for 10 years and, under the Directorship of Jenny Baker and Brian Dicks, was able to do much of the survey work that informs this volume. That work was mainly commissioned by the then Nature Conservancy Council (NCC) through their Head of Marine Conservation Branch, Roger Mitchell. My big break was to be taken on by the NCC in 1987 to lead the Marine Nature Conservation Review of Great Britain (MNCR), which finished in 1998. There are too many colleagues who were members of that MNCR team to thank individually, but what a good job they did! When the MNCR survey programme finished, I was greatly helped in finding a new role by one of English Nature's Directors, Sue Collins, and by their Head of Marine Conservation, Dan Laffoley. That new role was a secondment that became employment at the Marine Biological Association in Plymouth, where I had been discussing with its Director, Professor Mike Whitfield, a programme that would add value to what we knew about marine life by assessing its sensitivity to natural events and human activity: an idea that became the Marine Life Information Network (MarLIN). Mike took some brave decisions to jump-start a programme that had to find funding but, with the continued help of the next Director, Professor Steve Hawkins, MarLIN and its spin-outs has pulled together an enormous amount of information to underpin marine environmental protection, management and education. Again, there are too many colleagues to mention individually but what a good job they have done! Retirement beckoned, but there was too much still to do and I am grateful to the Council of the Marine Biological Association as well as its Director, Professor Colin Brownlee, and Deputy Directors, Professor David Sims, Matt Frost and Jon Parr for providing continued facilities at the Laboratory for me as an Associate Fellow.

Many colleagues and friends have helped me by reading through and commenting on draft sections of the book. Special thanks to Stewart Angus, Richard Baker, Brian Bett, Karen Boswarva, Blaise Bullimore, Peter Burgess, Roger Covey, Paul Dando, Matt Frost, John Howe, Dan Laffoley, Paula Lightfoot, Chris and Anne Mandry, Peter Messenger, Jon Moore, Nigel Mortimer, Paul Naylor, Nick Pope, Ivor Rees, Sue Scott, John and Karen Williams, and apologies to anyone not mentioned.

In 2016, I embarked on a tour to top up my underwater image collection at targeted locations. I am grateful to all those who helped me and especially those who provided hospitality: David and Jean Ainsley, Blaise and Lana Bullimore, Paul and Lucy Kay, Jon and Ginny Moore, Bill Sanderson, Sue and Mike Scott.

Scuba diving was of key importance in my working days and there are many team members who have maintained and developed the survey equipment that I used and ensured that health and safety regulations were adhered to. I was trained by the British Sub-Aqua Club (BSAC) and returned, on moving to Plymouth, to club diving with the Plymouth Sound Branch. Together with diving on Seasearch surveys, this book benefits greatly from my club diving – thanks to all in Plymouth Sound BSAC, especially for keeping me in my place. Others have helped get me into the water, especially Roy Lancaster and the crew of *Frolica* and Ilfracombe and North Devon branch of BSAC out of Ilfracombe, with John and Karen Williams from the *Blue RIB* and Peter Rowlands from MV *Magic* out of Plymouth.

I was pleased when Julie and Marc Dando (Wild Nature Press) agreed to take on publication of the book. It has been a very large task, undertaken with thoroughness and care to produce a volume that I am delighted with. Their work was assisted by Rowena Millar who undertook the editing to knock my text into shape.

Finally, to Bob Earll who reminded me that we should all write down everything we know before we die.

The author

Keith Hiscock is an ecologist who has been studying marine habitats and species for more than 50 years. A fortuitous combination of interests in marine biology, diving and photography and a great deal of good luck being in the right place at the right time has provided him with the opportunity to become a leading marine ecologist and conservation scientist in Britain. His early interest in seashore life and especially cold-water corals progressed through a degree in Zoology with Botany and then a PhD studying *The Influence of Water Movement on the Ecology of Sublittoral Rocky Areas*. After a short period monitoring rocky shores around Anglesey, he was appointed in 1975 as Deputy Director of the Field Studies Council Oil Pollution Research Unit at Orielton Field Centre in Pembrokeshire. From there, he, with many others, developed the methods and equipment that would be used to describe, catalogue and classify seabed habitats and their associated communities (now known as 'biotopes') around Britain through a series of studies commissioned by the then Nature Conservancy Council (NCC). When the NCC determined that a Marine Nature Conservation Review of Great Britain (MNCR) was needed, he was appointed to head that team in 1987. In 1998, the Review was 'finished', although far from complete. By now employed by English Nature, he had seen the need to bring together marine biological information and make it more useful for environmental protection and management, including through a new medium called 'The Internet'. That vision was achieved with the Marine Biological Association and is the Marine Life information Network (MarLIN) and its 'spin-outs': the Data Archive for Marine Species and Habitats and much of the education programme at the MBA. Having retired in 2007, he has continued to contribute to work at the Marine Biological Association and is an Associate Fellow there as well as pursuing those passions for marine biology, diving and photography in his own time.

The sponsors

Production of this book has been supported by the following organisations:

JNCC is the statutory adviser to the UK Government and devolved administrations on UK and international nature conservation. Its work contributes to maintaining and enriching biological diversity and sustaining natural systems. JNCC has a particular role in the UK's offshore waters by supporting government and industry in using our seas sustainably.

The Marine Biological Association promotes, supports and conducts scientific research into all aspects of life in the sea and works with its membership to provide a clear and independent voice on behalf of the marine biological community.

The Marine Conservation Society is the UK's leading charity working to protect our seas, shores and wildlife. The voice for our seas for almost 30 years, MCS focuses on protection for marine wildlife, sustainable fisheries and clean seas and beaches.

Natural England is the government's adviser for the natural environment in England, helping to protect England's nature, landscapes and coastal waters for people to enjoy and for the services they provide.

Now approaching its 30th year, **O'Three Ltd** has grown into an international brand renowned for its high end niche range of neoprene dry suits, wet suits, accessories and incredible detail for customer service. O'Three is passionate about the sea and is proud to have helped support *Exploring Britain's Hidden World*.

Since 1884, the **Scottish Association for Marine Science** has been dedicated to increasing, communicating and using our understanding of the global oceans for the public good through transformational research, inspirational education and public outreach: deploying our knowledge to solve real-world problems.

The Wildlife Trusts believe the best way to protect and manage our seas is to balance the needs of people and wildlife. They campaign for protected areas for wildlife at sea, run surveys around the coast and work with fishing industry and developers to reduce their environmental impact.

WWF is the world's largest leading independent conservation organisation. Their mission is to stop the degradation of our planet's natural environment, and build a future in which humans live in harmony with nature.